WORK BREAKDOWN STRUCTURES: THE FOUNDATION FOR PROJECT MANAGEMENT EXCELLENCE

Eric S. Norman, PMP, PgMP
Shelly A. Brotherton, PMP
Robert T. Fried, PMP

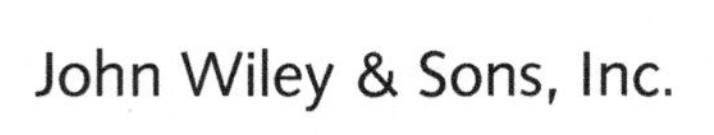

John Wiley & Sons, Inc.

Project Management Institute

Published by John Wiley & Sons, Inc., Hoboken, New Jersey.
Published simultaneously in Canada.

For general information on our other products and services please contact our Customer Care Department within the U.S. at (800)-762-2974, outside the U.S. at (317)-572-3993 or fax (317)-572-4002.

Project Management Institute (www.pmi.org) is the leading advocate for the project management profession globally. Founded in 1969, PMI has more than 400,000 members and credential holders in 174 countries. PMI's Project Management Professional (PMP) credential is globally recognized as the gold standard credential in project management

Wiley also publishes its books in a variety of electronic formats. Some content that appears in print may not be available in electronic books. For more information about Wiley products, visit our Web site at www.wiley.com.

Library of Congress Cataloging-in-Publication Data:

Norman, Eric S.
Work breakdown structures : The Foundation for Project Management Excellence / Eric S. Norman, Shelly A. Brotherton, Robert T. Fried.
p. cm.
Includes bibliographical references and index.
ISBN 978-0-470-17712-9 (cloth)
1. Project management. 2. Workflow. I. Brotherton, Shelly A. II. Fried, Robert T. III. Title.
HD69.P75N67 2008
658.4′04—dc22

10 9 8 7 6 5 4 3 2 1

Contents

Preface

So why write a book about Work Breakdown Structures? Actually, the answer is quite simple. In the vast body of project management literature, there is comparatively little written about Work Breakdown Structures (WBS)—and that is a problem. It is a problem because, not surprisingly, at this time in the evolution of Project Management as an advancing profession, the WBS has emerged as a foundational concept and tool. The WBS ensures clear definition and communication of project scope, while at the same time it performs a critical role as a monitoring and controlling tool. The WBS supports a variety of other Project Management processes—providing a baseline for planning, estimating, scheduling, and other "ing"-type activities. With the WBS performing this critical role, we believe it is important for Project Management literature to include additional material—and specifically, a detailed instructional text regarding Work Breakdown Structures.

This book is intended to be a beginning step in filling the information gap that exists between what is currently written about Work Breakdown Structures and what the authors have learned is needed by program and project managers today.

To be both specific and unscientific at the same time, if you were to stack all of the publications and writings about Work Breakdown Structures on a table, the pile would be an alarmingly scant 5 to 6 inches tall. We know this to be true—we've performed the experiment. Compare that with writings on risk management, project scheduling or the softer side of Project Management—negotiating, leading, managing projects, and so on, and the sky-scrapers of literature on these subjects dwarf the one-story dwelling of information on Work Breakdown Structures.

This book takes a new approach to the discussion of Work Breakdown Structures. First, you will notice that the text is laid out as a typical project might be managed—from concept and planning through delivery and project closeout, while the role of the WBS is explained during each step of the effort. Additionally, we use a single, common metaphor for the WBS throughout the text. While we discuss many applications and formats for the WBS, and we expand on concepts and examples throughout the book, this single metaphor provides a cohesive thread for you to follow from beginning to end.

Most importantly, however, the book is designed to function not only as a text, but also as a desk reference. To date, the vast majority of writings about Work Breakdown Structures have provided guidance about development of the tool or explain the benefits derived from using the WBS as a basis for planning.

Here, we explain how the WBS is first developed, then applied and continually referenced throughout the life of a project. So as a practitioner, if you are beginning a new project and are interested in examining how the WBS is utilized during Initiating and Planning phases, you will find helpful information in the early chapters of the book. By contrast, if you are at the midpoint in a project, perhaps during Executing or Monitoring and Controlling phases, the central chapters will provide reference material and information about the interplay between the WBS and other key Project Management processes such as Staff Planning, Communications, Risk, Change and Schedule Management. (*PMBOK® Guide*—Third Edition, pp. 121–122, 205–214)

Finally, we present new concepts for Work Breakdown Structures in this book. These new concepts relate specifically to what we will call *transition* activities. For example, woven throughout the chapters is the concept that during the life of a project, the WBS functions in different ways, depending on the phase of the project. For instance, during Initiating and Planning phases, the primary role of the WBS is to document and collect information, serving as a point of clarification that describes and defines—often in great detail—the boundaries of the project's scope as well as the "deliverables" and outcomes of the project. During Executing and Monitoring and Controlling phases, the WBS transitions from its passive role as a collection of information to one of action, in the role of project decision support, utilized as a reference and source for control and measurement. This key transition brings the WBS to life during the evolution of a project and is explained in the book.

To detail the layout of the book, the following is a brief synopsis that includes the three major parts and content of each chapter.

Part I, "Introduction to WBS Concepts," includes Chapters 1 and 2 of the book and is focused on the presentation of key baseline information regarding WBS constructs.

- In **Chapter 1** we present basic WBS concepts. This chapter provides background and foundational information on which the remainder of the book is based. Here, we include some general, some specific and a few very important definitions about Work Breakdown Structures. We also include a brief historic look at Work Breakdown Structures and introduce our key metaphor, the House example. More about this later.
- **Chapter 2** presents the WBS in more detail. In this chapter we discuss the key quality attributes of the WBS and explain the process of decomposition. We also present and explain a variety of WBS representations and describe how the quality attributes for a WBS apply similarly to projects, programs, portfolios, and ultimately, the enterprise.

Part II, "WBS Application in Projects," is the largest component of the book, comprising Chapters 3 through 10. It discusses the role of the WBS in each project phase from Initiating, through Planning, Executing, Monitoring, Controlling, and Closing. The chapters that make up the Planning discussions occupy the lion's share of Part Two—and that is a natural outcome, because the WBS is predominantly, though not exclusively, a planning tool. The House Metaphor will frequently appear in these chapters, and will take many forms. We will use this metaphor to ground our discussions and to provide a consistent and familiar place to return to when the going (and the detail of our writing) gets tough.

- **Chapter 3** discusses the role of the WBS in the Initiating phase and includes reviews of the interaction between the WBS and the Project Charter, the Preliminary Scope Statement, contracts, agreements, Statements of Work (SOW) and Contract Statements of Work, which apply to project components that have been contracted or subcontracted to providers outside the project or program organization.
- **Chapter 4** reviews how the *Product* Scope Description and the *Project* Scope Statement are used to build and construct a refined

Scope Baseline. In this chapter the initial construction of the Work Breakdown Structure occurs. Most importantly, in this chapter we explain the value of deliverable-based management as contrasted with an activity-based approach. Beyond this, key information and examples of the WBS Dictionary are presented along with examples of criteria for product and project acceptance.

- In **Chapter 5** we examine in detail the role of the WBS in build versus buy decisions along with in-depth discussions of the role of the WBS in Cost Estimating and Cost Budgeting.
- **Chapter 6** addresses key interactions of the WBS during additional planning activities. These include detailed discussions of the WBS during Quality Planning, Staff Planning, Risk Identification and Planning as well as the role of the WBS in the most carefully planned project management process—Communications Planning.
- **Chapter 7** introduces new concepts that enhance the process of translating the Scope Management aspects of the WBS into a representation of elements that can be used to create the Network Diagram. This process is examined in detail, and ultimately a completely new representation of scope—the Scope Relationship Diagram is presented. In this chapter, we also explain how the various individual WBS elements are oriented and related to one another. The Scope Relationship Diagram is introduced to represent how various elements of scope interact, and can in turn be used as a basis for Project Schedule development. While this book is not intended to detail the process of network diagramming or schedule development, the subjects are discussed briefly to help ease the transition from WBS to Project Schedule.

The role of the WBS in Executing and Monitoring and Controlling phases is discussed in Chapters 8 through 10. In these chapters, we discuss how the WBS performs the foundational role as a reference point for all project execution processes and is used as a source for decision making when risks or changes must be addressed. It should be noted here that we discuss Executing and Monitoring and Controlling activities together in this book. The Project Management Institute's *PMBOK® Guide*—Third Edition approaches this issue in two ways. In parts of Chapter 3 of the *PMBOK® Guide*—Third Edition, there is a clear separation between Executing and Monitoring and Controlling processes. A strong argument can be made, however, for describing the execution of a project as the work being performed within the effort while the Monitoring and Controlling process is simultaneously applied as the core Project Management activity that ensures the work is performed against the proper tasks, using the proper resources and is directed at the appropriate deliverables. The key theme, here, is that Executing and Monitoring and Controlling must be thought of as a common set of processes where Executing is the product-related activity and Monitoring and Controlling is performed as a guidance and management function over all other project activities. This concept is reinforced in the *PMBOK® Guide*—Third Edition. Figure 3.2 depicts how Monitoring and Controlling processes encompass and interact with all other project management process groups. The *PMBOK® Guide*—Third Edition defines this process: "The integrative nature of project management requires the Monitoring and Controlling process group interaction with every aspect of the other process groups" (*PMBOK® Guide*—Third Edition, p. 40).

As is the case in the Executing phase of a project and thereby, the Executing and Monitoring and Controlling portions of this book, Monitoring and Controlling activities are based largely on the scope definitions and detail provided by the WBS and the WBS Dictionary. Chapter 9 exposes the various functions of both the WBS and its associated Dictionary. These chapters should be truly seen as the "action" section of the book, representing the point in a project, program or portfolio where managers must manage and leaders must lead. If Initiating, Planning and Executing phases of a project (and by coincidence, the layout of this book) can be considered the setup discussion, then Monitoring and Controlling can be seen as the action part, with a strong emphasis on the controlling activities. It is between these phases that the role of the WBS is "transitioned" from passive to active.

- **Chapter 8** is the where the proverbial rubber meets the Project Management "road". In this phase of the project the WBS is transitioned from a planning tool to a decision support resource. The construction and design of this foundation tool is complete, and now the WBS is pressed into action as the source for decision guidance for the remainder of the project. In this chapter, we discuss how the WBS informs decisions that the Project Manager and stakeholders must make and aids in the process of Acquiring the Project Team (*PMBOK® Guide*—Third Edition, p. 209). Additionally, as the Project Manager leads and directs the project, the WBS guides

Project Execution and provides the starting point for Change Management (Scope Management) and Quality Assurance. Beyond this, the WBS and WBS Dictionary provide a foundation and baseline for Action and Issue, Risk, Change, Budget/Financial and finally Earned Value Management (EVM).

- **Chapter 9** examines how the WBS and WBS Dictionary are used as the foundation for decision making regarding project Performance Management and the essential activities the Project Manager must address regarding Stakeholder Management.

In the last section of Part II, we have a single chapter that discusses project closeout and carefully examines the activities the Project Manager must perform in order to successfully close, deliver and transition the completed project from the project team to the receiving organization or team. As in the Executing and Monitoring and Controlling project phases, this is a particularly active time for the WBS. Here it is used as a reference point to ensure all deliverables detailed in the WBS have been completed, delivered, accepted and turned over to the customer.

- **Chapter 10** includes a discussion of Scope Verification and Scope Management. These interactions within the project represent some of the most critically important activities of the project manager, program manager or leader, clearly articulating, communicating, reinforcing and protecting the boundaries established for the project. These boundaries are established through the Scope Statement, contracts and agreements that have been drafted and approved by the sponsor and key stakeholders. The WBS and WBS Dictionary are the key project documents/artifacts that represent these boundaries in a way that provides for the programmatic application of agreements within the project effort. This is the fundamental role of the WBS in a project management setting. Additionally in this chapter, we describe how customer acceptance criteria are referenced for each of the deliverables found in the WBS and described in the WBS Dictionary. Activities that signal completion are also included, such as training and warranty period coverage, support protocols, documentation turnover, contract closure, subcontract closure and formal sign-off and acknowledgment by the receiving organization.

The authors have included Part III of the book, "WBS for Project Management Decomposition," to help project managers clearly communicate the process components and outcomes that project managers routinely oversee during the performance of each project they lead. Though it is well understood that project managers perform an important, and perhaps critical, role in the delivery of projects, defining that role and the process deliverables associated with it has been a challenge for many. With this in mind, in Part III we discuss various ways of representing project management roles and deliverables that are an essential, often overlooked part of the project's work. Chapter 11 presents a focused discussion that describes various approaches for thoroughly and accurately representing project management in the WBS as an important component of the complete scope of the project.

- In **Chapter 11** we explain the origin of the WBS components we have included for Project Management and describe various approaches to the decomposition of the project management work present in nearly every project. In this chapter we provide a number of examples of the deliverables and outcomes that are seen as products of the project management process and share teachings about the best approach for specific projects. Most importantly, we have included two fully elaborated decompositions of the project management work as it is defined in The Project Management Institute's *PMBOK® Guide*—Third Edition. These two (complete) tables represent the views presented in the *PMBOK® Guide*—Third Edition and show the decomposition of project management deliverables by Process Group and by Knowledge Area as well. Your authors believe that the careful and accurate representation in the WBS of the project management work necessary to deliver a project is critical to the successful delivery of the project itself, and have provided this chapter to help ease that process for you.

This book presents three important themes. First is the idea that the WBS *transitions* from a paper/planning exercise during Initiating and Planning phases of the project to an *action* and *work performance* management tool in Executing, Monitoring and Controlling, and Closing. The second key theme is that a significant transition occurs between the scope defined by the WBS and the task, activity and milestone lists that make up the start of project schedule development. The third theme presented is that Monitoring and Controlling occurs throughout a project, from Initiating through Closing, and it is not an isolated process that occurs following Execution.

This book brings together the collective experience of three veteran, battle-hardened program/project managers. We met initially as Project Management Institute volunteers, where we made up three quarters of the project core team for the development of PMI's *Practice Standard for Work Breakdown Structures*—Second Edition. Although each of us has been blessed with a variety of experiences leading large and small initiatives with budgets ranging from thousands of dollars to those in the tens of millions, we each had faced similar challenges and expressed similar concerns about the clarity and concreteness of Project Scope and objective statements and our ability to manage to specific defined deliverables, once the projects we had been chartered to deliver got under way. This book describes our methods for clarifying, addressing and resolving those challenges, and provides us an opportunity to share our successes and lessons learned—big and small, with you.

We would like to thank our spouses, children, pets, colleagues, partners and associates who have supported us during the development of this book for their unending guidance, friendship, patience and love. Most importantly, one of our colleagues has had a significant impact on our writing—though he hasn't directly participated in the development of the book. We invoked his guidance and literate counsel as a regular part of our reviews and we are eternally grateful. George Ksander, we thank you for the countless contributions you've made to the quality of this effort. We would also like to extend a special thank you to Bob Argentieri and the folks at John Wiley & Sons, Inc for having faith in our vision—enough to encourage the development of the final product.

REFERENCES

Project Management Institute. 2004. *A Guide to the Project Management Body of Knowledge (PMBOK® Guide)—Third Edition.* Newtown Square, PA: Project Management Institute.

Project Management Institute. 2004. *Practice Standard for Work Breakdown Structures.* 2nd ed. Newtown Square, PA: Project Management Institute.

Foreword

The Work Breakdown Structure (WBS) is used as input for every other process of creating the schedule and the project budget. In that sense, the WBS is the foundation of the schedule and the budget as the foundation is for a house. If the foundation is weak the house will never be strong. It is hard to recover from a weak WBS. Needless to say, Eric and his team are focusing on a very important topic in this book. Rightfully they note that the available literature on this topic is thin. On the other hand, the WBS's we see in practice are often amazingly poor. This is not a paradox; the one explains the other perhaps. WBS's are an intuitive concept for some and an eternal mystery for others. It is time that some people wrap their mind around this issue. I think there is a real need for this book.

The book is like a handbook on how to implement *The Practice Standard for the WBS–2nd* Edition (published by PMI), another product of mostly this same team. These people were in fact so productive that the Practice Standard would have been a much thicker document were it not for the missive from PMI to keep the standard lean and mean. Not to worry, the team decided to simply publish all their brain products in this book as a separate publication.

The book builds entirely on top of the concepts in the PMBOK and the Practice Standard. It is valuable besides these standard documents in that:

- This book provides new processes that are the connecting dots between concepts explained on an abstract level in the standards. I liked for example the authors' new process of mapping out dependencies between WBS-deliverables, which I think is an essential activity, particularly in large programs.
- This book provides new concepts. For example, the notion that monitoring and controlling occur during the entire project, not just

during the PMBOK's *'Monitoring and Controlling'*. After all, who inspects the inspectors?

- This book provides new diagramming techniques. For example, the concept of defining relationships between scope elements cannot be found elsewhere. Other new things are the diagramming methods put forward by the authors: the *Scope Relationship Diagram* and the *Scope Dependency Plan*.

The authors take a courageous step by pushing the envelope for the betterment of our profession. I am personally very curious to see how these new processes, concepts and diagrams fare in the ocean of ideas and the often ferocious debates between thought leaders in our field.

There is no doubt in my mind that Eric Norman, Shelly Brotherton and Robert Fried have positioned themselves firmly in the category of project management thought leaders with this book. I encourage you to let them lead your mind for a few hours.

I hope you will enjoy reading this book as much as I did.

Eric Uyttewaal, PMP
Author of "Dynamic Scheduling with Microsoft Office Project 2003"
President of ProjectPro Corp.

Part I

Introduction To WBS Concepts

Chapter 1

Background and Key Concepts

"If you don't know where you're going, any road will take you there"

Anon

CHAPTER OVERVIEW

This chapter is placed up front not only because it is Chapter 1, but also because we wanted to provide background information for you before beginning the process of developing Work Breakdown Structures. This chapter introduces key concepts about the WBS that are discussed in much more detail later in the book, along with historic information about the emergence and evolution of the WBS over a number of decades. We also introduce the House metaphor.

The house what? The **House metaphor**. For our purposes, we will use the term metaphor here to mean a symbol or example that will represent how this concept can be applied in practice—although the example itself is fictitious. Actually, the House metaphor is a tool or rather, a section of a WBS from the construction of a house that we have developed for use throughout the book to help us illustrate our intended meaning—when words alone aren't enough to clarify and communicate key points or concepts. Following is an outline view of the House metaphor we will use, in one form or another, throughout the remainder of the text.

This metaphor is an important tool to cover at the beginning of the book because we will use it to describe, discuss and illuminate concepts throughout the text. We will use the House metaphor to illustrate examples, to provide a common, practical application of a topic or concept, and as a starting point for detailed examination of related topics.

1 House Project
  1.1 Primary Structure
    1.1.1 Foundation Development
      1.1.1.1 Layout—Topography
      1.1.1.2 Excavation
      1.1.1.3 Concrete Pour
    1.1.2 Exterior Wall Development
    1.1.3 Roof Development
  1.2 Electrical Infrastructure
  1.3 Plumbing Infrastructure
  1.4 Inside Wall Development: Rough Finish

Exhibit 1.1 House Metaphor—Outline Example

At the highest level, this chapter will contain the following:

- A general description of the Work Breakdown Structure and its role in project management
- WBS background and history
- Key terms and definitions
- The House Metaphor

WORK BREAKDOWN STRUCTURES

Let us begin...

Work Breakdown Structures were first used by the U.S. Department of Defense for the development of missile systems as far back as the mid-1960s, and they have been a fundamental component of the Project Management lexicon for nearly as long. The concept of the WBS and the practices around its use were initially developed by the U.S. Department of Defense (DoD) and National Aeronautics and Space Administration (NASA) for the purpose of planning and controlling large acquisition projects whose objective was development and delivery of weapons or space systems (Cleland, Air University Review, 1964, p. 14). These projects often involved many industrial contractors each with responsibility for separate components of the system and were managed by a central administrative office, either within a governmental agency or within one of the contracting firms which served as prime contractor. In this environment, the WBS was used to "...ensure that the total project is fully planned and that all derivative plans contribute directly to the desired objectives" (NASA, 1962).

The point is, that if true, and we assert right here that the statement is true, then the statement raises a question: "If the WBS is a fundamental building block for most projects, most of the time, then why are there so many conflicting viewpoints and approaches to development and use of Work Breakdown Structures?"

The answer to that question is somewhat elusive, and is one of the driving factors for writing this book. In the sections and chapters that follow we will examine various approaches to WBS development and will present a number of concepts, attributes, challenges and ultimately, recommendations for your consideration and use.

DEFINING WORK BREAKDOWN STRUCTURES

The *PMBOK® Guide*—Third Edition, defines a **Work Breakdown Structure** as "a deliverable-oriented hierarchical decomposition of the work to be executed by the project team to accomplish the project objectives and create the required deliverables. It organizes and defines the total scope of the project. Each descending level represents an increasingly detailed definition of the project work." The WBS is decomposed into Work Packages. **Work Packages** are defined in two different ways in the *PMBOK® Guide*—Third Edition. In the text, Work Packages are said to be the "lowest level in the WBS, and is the point at which the cost and schedule can be reliably estimated. The level of detail for Work Packages will vary with the size and complexity of the project. The **deliverable orientation** of the hierarchy includes both internal and external deliverables" (*PMBOK® Guide*—Third Edition, pp. 112, 114). Later in this chapter we provide the Work Package glossary definition for you.

There are a number of important concepts presented in this definition for the WBS. Of particular interest is the concept of deliverable orientation. The inclusion of these words is a key change from the definitions for the WBS in earlier editions of the *PMBOK® Guide* and reflects the expanded role the WBS performs in projects today. These changes are highlighted in Table 1.1.

Today, the WBS is understood to be more than an organization of the work of the project. The current definition, with the inclusion of the

Table 1.1 WBS Definition—Changes by Version

The Project Management Body of Knowledge (*PMBOK*®) (1987)	**A Guide to the Project Management Body of Knowledge (*PMBOK*® *Guide*) (1996)**	**A Guide to the Project Management Body of Knowledge (*PMBOK*® *Guide*—Second Edition) (2000)**	**A Guide to the Project Management Body of Knowledge (*PMBOK*® *Guide*—Third Edition) (2004)**
A task-oriented 'family tree' of activities.	A deliverable-oriented grouping of project elements which organizes and defines the total scope of the project. Each descending level represents an increasingly detailed definition of a project component. Project components may be products or services.	A deliverable-oriented grouping of project elements which organizes and defines the total scope of the project. Each descending level represents an increasingly detailed definition of a project component. Project components may be products or services.	A deliverable-oriented hierarchical decomposition of the work to be executed by the project team to accomplish the project objectives and create the required deliverables. It organizes and defines the total scope of the project. Each descending level represents an increasingly detailed definition of the project work. The WBS is decomposed into work packages. The deliverable orientation of the hierarchy includes both internal and external deliverables.

(**Sources:** Project Management Institute, *The Project Management Body of Knowledge (PMBOK®)*. PMI. Newtown Square: PA. 1987.; Project Management Institute, *A Guide to the Project Management Body of Knowledge (PMBOK® Guide)*. PMI. Newtown Square: PA. 1996.; Project Management Institute, *A Guide to the Project Management Body of Knowledge* (*PMBOK*® *Guide*—Second Edition) PMI. Newtown Square: PA. 2000.; Project Management Institute, *A Guide to the Project Management Body of Knowledge* (*PMBOK*® *Guide*—Third Edition) PMI. Newtown Square: PA. 2004.)

deliverable orientation wording, indicates that the process of developing the WBS includes the definition and articulation of specific outcomes of the project–the end products and results. By doing so, it becomes a reference point for all future project activities.

This critically important concept will be expanded later in the book, but we want to point to this definition as a departure point for our writing as well as a point of reference for you. Deliverable orientation is one of the Core Characteristics for the WBS, which we will discuss in Chapter 2. It is a fundamental attribute that will allow your WBS to be more than "shelfware" for your project, and will enable it to perform a critical role as a baseline document for communication of scope and outcomes during the initial phases of your project. In later phases, the WBS performs an active role as a basis for other key executing and monitoring and controlling activities. With these thoughts in mind, we can now take a broader look across the project management horizon to examine current trends and to establish context for our discussion.

There are additional reasons for preferring a deliverable orientation for WBS construction over task/activity or process orientations. With process and task-oriented Work Breakdown Structures, the deliverables or outcomes described by the WBS are the project processes themselves, rather than the project's products or outcomes. When this is the case, the project team spends a great deal of energy on refinement and execution of the project's processes, which can ultimately become models of care and efficiency–but that do not necessarily produce the desired outcomes for the project because the focus has been on the process of producing outcomes, not the outcomes themselves.

Additionally, task/activity WBS construction is truly a contradictory concept from the outset. As we will examine later, tasks and activities are truly part of the project scheduling process and have no place in the WBS to begin with. Later, in Chapter 7, we will discuss the creation of the Project Schedule and will explain that tasks, activities and milestones are outcomes of the decomposition of the WBS that extends beyond the Work Package level, (the lowest level of decomposition of the WBS) and yields elements that are carried forward into the project schedule. So from our perspective, developing a WBS based around tasks and activities is simply a contradiction in terms. To us, and to those who wish to develop high-quality Work Breakdown Structures that focus attention on outcomes and deliverables, this truly cannot be useful.

IMPORTANCE OF THE WBS

Everyday practice is revealing with increasing regularity that creation of a WBS to define the scope of the project will help ensure delivery of the project's objectives and outcomes. There are numerous writings that point to the WBS as the beginning step for defining the project and insist that the more clearly the scope of the project is articulated before the actual work begins, the more likely the success of the project. Here are a few examples from recognized, reliable Project Management sources:

- John L. Homer and Paul D. Gunn "The intelligent structure of work breakdowns is a precursor to effective project management." (Homer and Gunn, 1995, p. 84).
- Dr. Harold Kerzner: "The WBS provides the framework on which costs, time, and schedule/performance can be compared against the budget for each level of the WBS" (Kerzner 1997, p. 791).
- Carl L. Pritchard: "The WBS serves as the framework for project plan development. Much like the frame of a house, it supports all basic components as they are developed and built" (Pritchard 1998, p. 2).
- Dr. Gregory T. Haugan: "The WBS is the key tool used to assist the project manager in defining the work to be performed to meet the objectives of a project" (Haugan, 2002, p. 15).
- The *PMBOK® Guide*—Third Edition stresses the importance of the WBS in the Planning Process Group, which begins with three essential steps—Scope Planning (3.2.2.2), Scope Definition (3.2.2.3) and Work Breakdown Structure Development (3.2.2.4). (*PMBOK® Guide*—Third Edition).

Experienced Project Managers know there are many things that can go wrong in projects regardless of how successfully they plan and execute their work. Component or full-project failures, when they do occur, can often be traced to a poorly developed or nonexistent WBS.

A poorly constructed WBS can result in negative project outcomes including ongoing, repeated project re-plans and extensions, unclear work assignments for project participants, scope creep, and its sister, unmanageable, frequently changing scope, as well as budget overruns, missed deadlines and ultimately unusable new products or delivered features that do not satisfy the customer nor the objectives for which the project was initiated.

WBS LESSON LEARNED: A BRIEF ILLUSTRATION

Why is this the case? How can all of these problems be linked to the completeness or quality of the WBS? To answer this question, let us take a brief look at what typically happens following missed deliverables or project component failures. Once it becomes obvious something will be missed by the project team—the delivery date, key features or functionality, or perhaps the budget, the dust settles.

Shortly afterward (and exactly how long "shortly" is can vary quite a bit) the project leader and functional managers stop looking for someone to blame and cooler heads prevail. Quite often someone emerges (most likely an executive or Project Sponsor) and asks to see the "project's documentation." At this point the Project Manager scrambles to produce the project plan, project schedule, risk plan and register, change request log and the WBS for the project—if it exists. In a very short time, this person, who hasn't been close to the project on a day-to-day level, "down in the trenches" with the project team, will undoubtedly pull out a single project document and point to specific wording that describes precisely what should have been delivered by the project team, and when. That document is often the Scope Statement, the project's Charter or its contracts and agreements.

Having found the desired scope statements and agreements, the project executive or sponsor will call a series of meetings with the appropriate responsible parties, and will ask some very pointed questions about the reasons the project didn't result in the outcomes specified in the foundational documents—and will immediately begin negotiations to get what he/she intended to have delivered, delivered. Most notably, the project executive or sponsor may, at this point commit to ensuring the delivery will happen by taking a much more active role in the day-to-day activities of the project. This is *not* the most desirable outcome for a Project Manager wishing to be the master of his or her own project destiny.

Examining this scenario a little more closely, we can find the root cause. The sponsor/executive wants to take a more active role in ensuring the project has a higher likelihood of reaching its desired objectives because he or she believes that key project information, vital to making decisions about the project's outcomes didn't reach the decision maker(s). Clearly, this was a communications problem from the beginning. It truly doesn't matter whether the Project Manager believes the project communications were effective or not. The sponsor/executive believes they were not, and is taking an active role as a result. Key deliverables were missed—and there had been plenty of opportunity to surface the issues relating to the absent scope elements.

So what can the Project Manager do to learn from this experience? Beyond learning how to manage the pain of embarrassment and lead the recovery process following the missed deliverable(s), the Project Manager should look carefully at root causes. So now would be a good time for the Project Manager to ask him or herself, "What is (frequently) the cause for this scenario?" The answer is fairly straightforward: poor communication and validation of changes to the approved scope, schedule, and feature/functionality.

When this occurs, the Project Manager very quickly realizes that the obvious solution to the problem exists within the project's documentation.

Had the project's WBS clearly articulated the project deliverables (internal, interim and endpoint) and outcomes, at each critical interval along the way to delivery, the Project Manager could have validated progress against the stated scope–represented by the WBS. When challenges to scope and schedule were presented to the sponsor and/or other stakeholders, using the Change Management process for the project, these could be balanced against the documented, agreed-upon scope and feature/functionality described by the WBS and explained by the project plan. In the absence of clear WBS deliverables and outcomes, these discussions are considerably more elusive and difficult.

For the Project Manager, it's a lesson learned. For this discussion, the scenario becomes a template for defining critical success factors for scope management and communication. Those factors include a clearly articulated WBS, a scope management and scope control process (Change Management), and an effective communications process that will enable the Project Manager to articulate agreed-upon deliverables and the decisions that affect the schedule for completion of those deliverables.

It is essential for the Project Manager to find tools that will help communicate the frequency and impact of changes that follow the initiating and planning phases of the project–when the WBS is finalized and approved. If the WBS for the project was constructed so that it clearly defined the deliverables and outcomes for the project–including those that are transitional or temporary (interim) in nature, prepared for internal organizations as well as the end customer, then the Project Manager has at his or her fingertips a highly valuable tool. The WBS becomes the static document that can be referenced in an unemotional manner.

To avoid these project pitfalls, the WBS is used as a foundational building block for the initiating, planning, executing, and monitoring and controlling processes and is central to the management of projects as they are described in the *PMBOK® Guide*—Third Edition. Typical examples of the contribution the WBS makes to other processes are described and elaborated in the *Practice Standard for Work Breakdown Structures*–Second Edition.

To explain, there are many project management tools and techniques that use the WBS or its components as input (*PMBOK® Guide*—Third Edition, Chapter 5, Section 5.3). For example, the WBS utilizes the **Project Charter** as its starting point. The high-level elements in the WBS should match, as closely as possible, the nouns used to describe the outcomes of the project in the **Project Scope Statement**. In addition, the **Resource Breakdown Structure** (RBS) describes the project's resource organization and can be used in conjunction with the WBS to define work package assignments. The **WBS Dictionary** defines, details, and clarifies the various elements of the WBS.

Transitioning from the WBS to the Project Schedule is discussed in Chapter 7 and takes a number of references from the chapter on Project Time Management of the *PMBOK® Guide*—Third Edition. **Activity Definition**, the starting point for project schedule development relies on the WBS for the decomposition process, beginning at the lowest level of the WBS–the Work Package, to produce relevant project tasks, activities and milestones. Activity Sequencing describes and illustrates the logical relationships between these tasks, activities and milestones and shows the dependencies and precedence for each, orienting them in a Project Schedule Network Diagram.

Whether you choose Arrow Diagram Method (ADM), where the activities are shown on arrows that link nodes of the network diagram (Activity On Arrow), or the Precedence Diagram Method (PDM) where the nodes represent the project's activities while the arrows depict dependencies between them (Activity On Node), the starting point for this process is the WBS, where the scope of the project has been carefully decomposed to the Work Package level.

The WBS is also used as a starting point for **Scope Management** and is integral to other Project Management processes, and as a result, the standards that define these processes explicitly or implicitly rely on the WBS. Standards that take advantage of the WBS either use the WBS as an input (e.g., PMI's *Practice Standard for Earned Value Management (EVM)* and the *Practice Standard for Scheduling* or incorporate the WBS as the preferred tool to develop the scope definition (e.g., the *PMBOK®*

Guide—Third Edition, *OPM3*®). Beyond this, other practices recognized world-wide frequently reference the WBS as the starting point for scope. These practices include Prince2, (Projects in Controlled Environments), CMMI (Capability Maturity Model Integration) and RUP, the (Rational Unified Process).

WBS CONCEPTS

As we noted at the beginning of this chapter, the WBS, as defined in the *PMBOK*® *Guide*—Third Edition, is "a deliverable-oriented hierarchical decomposition of the work to be executed by the project team to accomplish the project objectives and create the required deliverables. It organizes and defines the total scope of the project. Each descending level represents an increasingly detailed definition of the project work. The WBS is decomposed into work packages."

With this definition, it is clear the WBS provides an unambiguous statement of the objectives and deliverables of the work performed. It represents an explicit description of the project's scope, deliverables and outcomes—the "what" of the project. The WBS is not a description of the processes followed to perform the project... nor does it address the schedule that defines how or when the deliverables will be produced. Rather, the WBS is specifically limited to describing and detailing the project's outcomes or scope. The WBS is a foundational project management component, and as such is a critical input to other project management processes and deliverables such as activity definitions, project network diagrams, project and program schedules, performance reports, risk analysis and response, control tools or project organization.

DESCRIBING THE WBS

The upper levels of the WBS typically reflect the major deliverable work areas of the project, decomposed into logical groupings of work. The content of the upper levels can vary, depending on the type of project and industry involved. The lower WBS elements provide appropriate detail and focus for support of project management processes, such as schedule development, cost estimating, resource allocation, and risk assessment. The lowest-level WBS components are called, as we've discussed earlier, Work Packages. The glossary definition for Work Package is, "A deliverable or project work component at the lowest level of the Work

Breakdown Structure. The work package includes the schedule activities and schedule milestones required to complete the work package deliverable or project work component" (*PMBOK® Guide*—Third Edition, p. 380). These Work Packages define and contain the work to be performed and tracked. These can be later used as input to the scheduling process to support the elaboration of tasks, activities, resources and milestones which can be cost estimated, scheduled, monitored, and controlled.

Here are a few of the key characteristics of high-quality Work Breakdown Structures (*Practice Standard for Work Breakdown Structures*–Second Edition):

- A central attribute of the WBS is that it is "deliverable oriented" (Berg and Colenso (2000)). The *PMBOK® Guide*—Third Edition defines a deliverable as "Any unique and verifiable product, result, or capability to perform a service that must be produced to complete a process, phase or project." In this context, *oriented* means aligned or positioned with respect to deliverables (i.e., focused on deliverables).
- An additional key attribute of the WBS is that it is a "... hierarchical decomposition of the work..." Decomposition is "a planning technique that subdivides the project scope and project deliverables into smaller, more manageable components, until the project work associated with accomplishing the project scope and deliverables is defined in sufficient detail to support executing, monitoring, and controlling the work" (*PMBOK® Guide*—Third Edition, p. 358). This decomposition (or subdivision) clearly and comprehensively defines the scope of the project in terms of individual subdeliverables that the project participants can easily understand. The specific number of levels defined and elaborated for a specific project should be appropriate for effectively managing the work in question.
- The 100% Rule (Haugan, 2002, p. 17) is one of the most important principles guiding the development, decomposition, and evaluation of the WBS. This rule states that the WBS includes 100% of the work defined by the project scope and, by doing so, captures *all* deliverables—internal, external and interim—in terms of work to be completed, including project management. The rule applies at all levels within the hierarchy; the sum of the work at the "child" level must equal 100% of the work represented by the "parent". The WBS should not include any work that falls outside the actual scope of the project; that is, it cannot include more than 100% of the work.

THE HOUSE METAPHOR–A CONSISTENT EXAMPLE

Throughout the book, and from this point forward, we will be discussing the use of the WBS in the execution of projects. We will show how the WBS is designed and created during the Initiating and Planning phases, and we will describe and illuminate the ways in which the WBS is used as a basis for decision making throughout the remainder of the project, during project Executing, Monitoring and Controlling, and Closing.

With this in mind, we have developed a small example–a fictitious project WBS that we will refer to when we want to explain a concept or describe the application of a theory. This example–our House Construction WBS, is intentionally sparse and not precisely correct. We are using it as a metaphor for other, more complete Work Breakdown Structures.

Here, the House Metaphor will allow us to communicate key concepts and help articulate our dialog with you. When it becomes necessary, and to reinforce concepts we present in later chapters, we will use other WBS examples that are fully elaborated and complete. But when we do that, we will be reinforcing concepts we have presented previously, or showing how more than one concept is linked together in a larger WBS example.

As we begin, however, we'll have to ask you to join us in "suspending the disbelief" about the accuracy and design of the House Metaphor. If you are familiar with home construction or work in the construction industry, it is likely you will find plenty of reasons to challenge our example. In fact, if you are familiar with the development of Work Breakdown Structures, you will also likely find lots of opportunity to dissect the House Metaphor. For you to find value in the book, we want you to accept the House Metaphor as valid and agree to accept it as the example it is for describing concepts.

The House Metaphor in its most simple construction–the outline view is shown in Exhibit 1.2.

You will see this example many times throughout the book. We will use it in its complete form, we'll take excerpts from it, we'll represent it in other ways and we'll show you how this example relates to other project management processes by elaborating various parts of the House Metaphor. Whatever the case, we'll use this example as a thread, or series of breadcrumbs we will leave for you so you can always find the path through the book. Look for the House Metaphor, and just like that, you'll be back on the right track.

1 House Project
  1.1 Primary Structure
    1.1.1 Foundation Development
      1.1.1.1 Layout—Topography
      1.1.1.2 Excavation
      1.1.1.3 Concrete Pour
    1.1.2 Exterior Wall Development
    1.1.3 Roof Development
  1.2 Electrical Infrastructure
  1.3 Plumbing Infrastructure
  1.4 Inside Wall Development: Rough Finish

Exhibit 1.2 House Metaphor—Outline Example

CHAPTER SUMMARY

This chapter presents key topics regarding the history of the WBS and its application in projects, from early use and development with the U.S. Department of Defense and NASA (National Aeronautics and Space Administration) to the current application of Work Breakdown Structures in projects today.

Most importantly, this chapter introduces a number of fundamental truths about Work Breakdown Structures, including the significant evolution of the definition itself. This striking evolution shows how thinking about the WBS has progressed and advanced over the past forty years, from what was then a simple statement about its attributes . . .

- A "task-oriented family tree of activities"

to what is now recognized internationally as the latest thinking about it's utility and function,

- "A deliverable-oriented hierarchical decomposition of the work to be executed by the project team to accomplish the project objectives and create the required deliverables. It organizes and defines the total scope of the project. Each descending level represents an increasingly detailed definition of the project work. The WBS is decomposed into work packages. The deliverable orientation of the hierarchy includes both internal and external deliverables."

This detailed and fully elaborated definition reflects the care that has been taken over the past four decades to explain and document the true function and role the WBS performs as a foundational building block for projects and programs.

Basic concepts, essential to understanding and effectively applying Work Breakdown Structures are presented in this chapter. First, we explain how each descending level of the WBS is accomplished through a process of decomposition to reach the lowest level of the WBS, the Work Package. We additionally introduce the 100% Rule guidance provided from a highly regarded authority on Work Breakdown Structures from the U.S. military. This concept along with a host of others is briefly discussed in this chapter. In later chapters, each of the concepts is elaborated in greater detail.

Finally, this chapter introduces a new concept and a few breadcrumbs to aid in the journey through WBS principles and practice. This series of breadcrumbs starts with the House Metaphor a fictitious example the authors have developed to establish a common path and theme throughout the book from Initiating and Planning through Executing, Monitoring and Controlling and finally Closing. The concepts presented throughout the book utilize this metaphor and rely on its simplicity to help guide the reader through a typical project lifecycle starting in this chapter and progressing through Chapter 10.

REFERENCES

Berg, Cindy, and Kim Colenso. 2000. Work breakdown structure practice standard project—WBS vs. activities. *PM Network* 14(4) (April): 69–71.

Cleland, David I. 1964. Project Management: An Innovation of Thought and Theory. *Air University Review* (November–December): 16.

Haugan, Gregory T. 2002. *Effective Work Breakdown Structures*. Vienna, VA Management Concepts.

Homer, John L., and Paul D. Gunn. 1995. *Work structuring for effective project management*. Project Management Institute 26th Annual Seminar/Symposium. New Orleans, Louisiana, October 1995, p. 84.

Kerzner H. 1997. *Project management: A systems approach to planning, scheduling, and controlling*. 6th ed. New York: John Wiley & Sons.

National Aeronautics and Space Administration. *NASA PERT and Companion Cost System Handbook*. Washington, DC. U.S. Government Printing Office, 1962.

Pritchard, Carl L. 1998. *How to Build a Work Breakdown Structure: The Cornerstone of Project Management.* Arlington VA. ESI International.

Project Management Institute. 1987. *The Project Management Body of Knowledge (PMBOK®).* Newtown Square, PA: Project Management Institute.

Project Management Institute. 1996. *A Guide to the Project Management Body of Knowledge* Newtown Square, PA: Project Management Institute.

Project Management Institute. 2000. *A Guide to the Project Management Body of Knowledge (PMBOK® Guide—)* Second edition Newtown Square, PA: Project Management Institute.

Project Management Institute. 2003. *Organizational Project Management Maturity Model (OPM3®).* Knowledge Foundation. Newtown Square, PA: Project Management Institute.

Project Management Institute. 2004. *A Guide to the Project Management Body of Knowledge (PMBOK® Guide)*—Third edition. Newtown Square, PA: Project Management Institute.

Project Management Institute. 2004. *Practice Standard for Earned Value Management (EVM).* Newtown Square, PA: Project Management Institute.

Project Management Institute. 2004. *Practice Standard for Work Breakdown Structures*—Second edition. Newtown Square, PA: Project Management Institute.

Project Management Institute. 2007. *Practice standard for Scheduling.* Newtown Square, PA: Project Management Institute.

CHAPTER QUESTIONS

1. According to current PMI standards, Work Breakdown Structures are:
   a. Task-oriented
   b. Process-oriented
   c. Deliverable-oriented
   d. Time-oriented
2. The elements at lowest level of the WBS are called ______________.
   a. Control Accounts
   b. Work Packages
   c. WBS deliverables
   d. Lowest-level WBS elements
3. The ______________ is utilized as the starting point for creating a WBS.
   a. Preliminary Project Scope Statement
   b. Product Scope Description
   c. Final Product Scope Statement
   d. Project Charter
4. Which of the following are key characteristics of high-quality Work Breakdown Structures? (Select all that apply.)
   a. Task-oriented
   b. Deliverable-oriented
   c. Hierarchical
   d. Includes only the end products, services or results of the project
   e. Completely applies the 100% Rule
5. Who initially developed Work Breakdown Structures?
   a. U.S. Department of Defense and NASA
   b. Builders of the great pyramids of Egypt
   c. Architects of the Roman Coliseum
   d. Russian Space Agency

Chapter 2

Applying WBS Attributes and Concepts

CHAPTER OVERVIEW

This chapter provides an introduction to key WBS concepts and describes the characteristics of WBS quality. WBS attributes are described here, as is the concept of decomposition. The chapter goes on to examine the use of the WBS as it applies to projects and beyond — to programs, portfolios and the enterprise or institutional level as well. The chapter concludes with a discussion of WBS representations as well as tools for creating and managing Work Breakdown Structures.

The major sections of this chapter are:

- WBS Attributes
- Decomposition
- WBS in Projects, Programs, Portfolios and the Enterprise
- WBS Representations
- WBS Tools

WBS ATTRIBUTES

While there is general agreement that the WBS is the foundation upon which a great many project management processes and tools are based, there is surprisingly little agreement on the best method for creating the WBS. So before we delve into the creation and usage of the WBS in the project management life cycle in the coming chapters, it is important to understand the key attributes of a WBS. By describing the attributes of a high-quality Work Breakdown Structure, you can begin to understand the true power of the WBS and how to effectively apply it to projects, programs, portfolios and beyond.

With the release of the Project Management Institute's (PMI) *Practice Standard for Work Breakdown Structures*—Second Edition, two quality principles for the WBS were introduced. These quality principles accurately capture the essence of a high-quality WBS:

1. A quality WBS is a WBS constructed in such a way that it satisfies all of the requirements for its use in a project (*Practice Standard for Work Breakdown Structures*—Second Edition, p. 19),
2. WBS quality characteristics apply at all levels of scope definition (*Practice Standard for Work Breakdown Structures*—Second Edition, p. 22).

From these two principles, it becomes clear that a quality WBS must satisfy all of the requirements for its use, at any level, be it project, program, portfolio or enterprise. This concept of fulfilling use requirements also meshes perfectly with the concept of quality as defined in the *PMBOK® Guide*—Third Edition. There quality is defined as "the degree to which a set of inherent characteristics fulfills requirements" (*PMBOK® Guide*—Third Edition, p. 371).

When applying these quality principles to the real world, it quickly becomes apparent that *use* is sometimes relative to the project or program to which it is being applied. The *Practice Standard for Work Breakdown Structures*—Second Edition takes this into account by further defining two subprinciples. These quality subprinciples introduce the concepts of Core and Use-Related Characteristics.

WBS Core Characteristics

Core Characteristics are the *minimum* set of specific attributes that must be present in every WBS. If a WBS adheres to these characteristics, it is said to have Core Quality. These Core attributes are very black or white and contain no shades of gray. A WBS either possesses these Core Characteristics or it does not. A WBS with Core Quality:

- Is deliverable-oriented
- Is hierarchical and constructed in such a manner that (a) each level of decomposition includes 100% of the work of its parent element, and (b) each parent element has at least two child elements

- Defines the full scope of the project and includes all project related work elements including all internal, external and interim deliverables
- Includes only those elements to be delivered by the project (and nothing that is considered out of scope)
- Uses nouns and adjectives to describe the deliverables, not verbs
- Employs a coding scheme that clearly depicts the hierarchical nature of the project
- Contains at least two levels of decomposition
- Is created by those performing the work with technical input from knowledgeable subject matter experts and other project stakeholders
- Includes Project or Program Management at level 2 of the hierarchy
- Includes a WBS Dictionary that describes and defines the boundaries of the WBS elements
- Contains work packages that clearly support the identification of the tasks, activities and milestones that must be performed in order to deliver the work package
- Communicates the project scope to all stakeholders
- Is updated in accordance with project change management procedures

The deliverable orientation of a WBS is so important that it is included directly in the definition of the Work Breakdown Structure. Deliverables, as the work products of the project, allow for the total scope of the project to more easily be defined.

As noted in Chapter 1, the WBS defines the "what" of the project. Each WBS element must be expressed as a noun and adjective, not verb and object, which would imply action. The noun and adjective form ensures that the WBS elements are expressed as deliverables, not tasks. The example in Figure 2.1 depicts a quality deliverable-oriented WBS. Note how the deliverables, at each level, are labeled using nouns and adjectives. Now contrast the example in Figure 2.1 with the task-oriented illustration in Figure 2.2.

In this example, WBS elements are expressed using the verb-object form. In these types of process or task-oriented Work Breakdown Structures, the work is described as a process or action. This implies that the true end-product of the WBS element is a refined process rather than the actual deliverables of the process. The performance of the process,

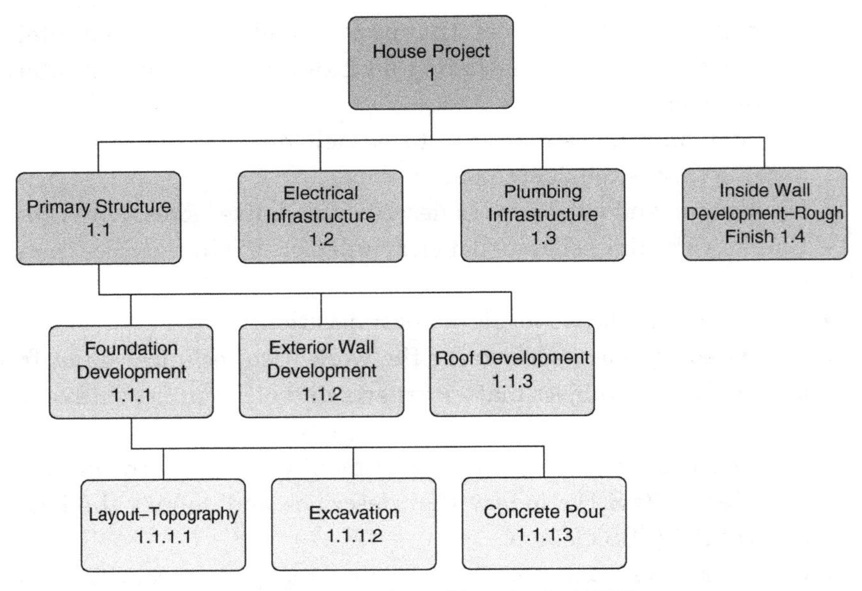

Figure 2.1 Deliverable-oriented WBS.

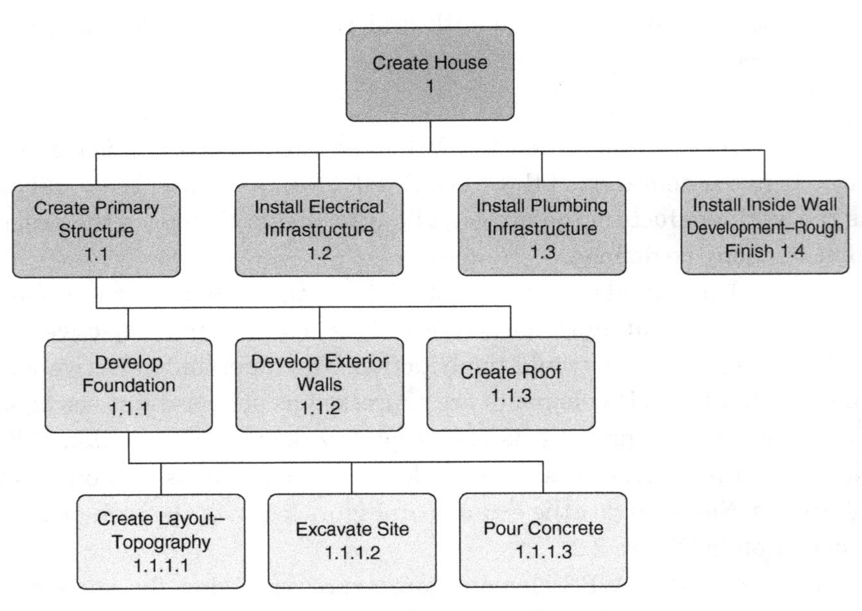

Figure 2.2 Task-oriented WBS.

rather than outputs, becomes the focus of the work, and as such it is possible to perform the process flawlessly without ever producing specific deliverables. If this is the case, it becomes increasingly difficult to know when deliverables have been completed or when acceptance criteria have been met. At face value, process-oriented Work Breakdown Structures appear logical and complete, but also serve to cloud or obscure the true objectives (outcomes) of the work. Issues that can be directly related to the development of process-oriented Work Breakdown Structures include lack of clear definition of project deliverables and infrequent use of the WBS Dictionary to define and explain the outcomes. Frequently absent, also, are boundaries and completion criteria for each deliverable. This results in difficulty measuring deliverable progress, as portions of project components may be spread across multiple WBS elements in the hierarchy (Pritchard 1998, p. 9). For these reasons, Figure 2.2 does not represent a quality WBS.

Another important aspect of these WBS Core Characteristics is the reinforcement of the idea that the WBS includes the total scope of the project effort, including all deliverables. By the same token, the WBS should absolutely not contain any element or deliverable that is not included in the project's scope. Many project managers often rely on the old saying, "If it's not in the WBS, it's not in scope." This is as true today as it ever was.

The Core Characteristics also introduce the concept of the **100% Rule**. This rule states that "the next level of decomposition of a WBS element (child level) must represent 100 percent of the work applicable to the next higher (parent) element" (Haugan p. 17). This rule ensures that as the WBS hierarchy is decomposed into additional levels of detail, no component of scope is lost. By ensuring that each level of decomposition includes 100% of the parent elements deliverables, a Project Manager can ensure that the end result of the WBS definition is a set of work packages that completely defines all the deliverables required by the project or program. The 100% Rule is fundamental to the development of Work Breakdown Structures. We will reference it many times throughout the remainder of this book.

These Core Characteristics also describe basic attributes that should be present in every WBS, regardless of the project type, industry, or context. All work breakdown structures should be hierarchically defined with a coding scheme that clearly illustrates the WBS hierarchy. This is necessary to ensure there is a close and easily recognizable relationship between all of the elements in a particular WBS, while simultaneously making it clear that there is a distinction between the WBS elements from one project and those from a different project's WBS. To illustrate, imagine two projects from a program called the "Red Program". The two projects in the Red Program are Project A and Project B. Three WBS elements from Project A are called A.1—Test Planning; A.1.1—Expected Results and A.1.2—Test Cases. Project B also employs precisely the same WBS elements, but names them uniquely for their use. Project B defines these same WBS elements as B.1—Test Planning; B.1.1—Expected Results and B.1.2—Test Cases. Although these elements produce the same outcomes, use the same words and are used in two projects within the same program, they are easily identifiable as parts of separate projects.

In Figures 2.1 and 2.2, the top of the chart reflects the project level and is coded with a 1. The next level of decomposition, level 2, is broken down into 1.1, 1.2, 1.3 and 1.4. This type of hierarchical coding scheme is a critical element to quality Work Breakdown Structures. By employing a unique coding scheme for each WBS element in a project, including the WBS element at level 1, it is easy to identify distinctions between one project and another, one grouping of elements and another or single WBS elements, even if the WBS elements are presented a single, standalone quantities. For instance, it is easy to see that a WBS element hierarchically labeled "X.2.1.1" is not part of the WBS structure that includes the hierarchy labeled "A.B.3.2.1." The first is a WBS element from project X, while the latter is an element from project A. As you can see, the coding scheme need not be all numeric. In fact, any combination of characters or icons, so long as they are hierarchical, will work just fine.

Although a WBS is hierarchical, that does not mean that the representation of the WBS must be depicted as a hierarchical, organizational-type graphic diagram. This may be a very familiar representation of the WBS, but a quality WBS can be represented in many forms, as you will see later in this chapter—all of which are valid. The important aspect is to ensure that the WBS clearly communicates the project scope to all of the project's stakeholders. To accomplish this effectively, a project manager may choose to illustrate the WBS in more than one way.

Core Characteristics for the WBS include the concept of Work Packages, the lowest level WBS element. This concept is important because it is these Work Packages that will be further decomposed into tasks, activities and milestones as part of the transition from the WBS to the Project Schedule. Although this transition is described in more detail in later chapters, it is important to note here that the Work Packages must clearly support the identification of the tasks, activities and milestones that will be required to properly create the intended project deliverables.

WBS Use-Related Characteristics

If it is true that a quality WBS is constructed in such a way that it satisfies all of its intended needs, then what is it about those needs that differentiate one WBS from another? We are certain that all Work Breakdown Structures are not the same—so what then, separates and makes them unique?

This is where the concept of **Use-Related Characteristics** comes into play. Use-Related Characteristics include those additional attributes that vary from one project to next, across industries, environments or in the way the WBS is applied within the project. With Use-Related Characteristics, the quality of the WBS depends on how well the specific content and types of WBS elements address the full set of needs of the project or program. This then implies that the more needs that the WBS meets, the higher the resulting quality of the WBS.

Examples of Use-Related Characteristics include, but are not limited to, the following:

- Achieves a sufficient level of decomposition to enable appropriate management and control
- Provides sufficient detail for bounding and communicating the scope of the project in its entirety
- Contains specific types of WBS elements necessary for the project
- Clearly enables the assignment of accountability at the appropriate level, regardless of whether the WBS is for a program or an individual project

So what, exactly, does *sufficient* mean in these examples? Well, like many other answers in project management, the real answer is, "It depends." As we noted earlier, the concept of Use-Related Characteristics implies that the answer to that question differs from project to project. The needs of one project differ greatly from the needs of the next. Given this, *sufficient* is the degree that a particular need is fulfilled for use on the specific project.

No two projects and programs are ever the same. Differences can be the result of project type, organization, type of deliverables, complexity or any number of other factors. Given the differences between projects, there is no one standard that can be applied to all projects for the level of decomposition required to enable the appropriate management and control for the project. Complex projects may require additional levels of decomposition when compared to a simple project. This is also true when communicating project scope. The level of detail required to effectively bound and communicate project scope is also dependent on the requirements of the project in question.

As all projects are different, so too are their requirements. This implies that the types of WBS elements required to satisfy those requirements will be different from project to project. Some projects may require level-of-effort WBS elements while others may require discrete elements based on a specific product development life cycle. **Level-of-effort** is defined by PMI as any "support-type activity (e.g., seller or customer liaison, project cost accounting, project management, etc.), which does not produce definitive end products. It is generally characterized by a uniform rate of work performance over a period of time determined by the activities supported" (Errata of the *PMBOK® Guide*—Third Edition, p. 363). This differs from discrete effort, which is defined as "work effort that is separate, distinct, and related to the completion of specific work breakdown structure components and deliverables, and that can be directly planned and measured" (Errata of the *PMBOK® Guide*—Third Edition, p. 359). As the definition notes, level-of-effort WBS elements are generally used in instances where there is no concrete deliverable. Examples of this include administrative project tracking and actual project management and control. Discrete WBS elements are those where there is a tangible, definable deliverable. Examples of this can be a document, a widget or even a specific result (tested application).

The examples in Figure 2.3 and 2.4 are illustrations taken from two Work Breakdown Structures. They are included here to show how the WBS level of detail will differ, depending on the needs of the project. While one WBS is very brief and simple, the other is larger and more complex. Both are valid Work Breakdown Structures, but the efforts they represent have different intents and outcomes and represent vastly differing scope. Figure 2.3 can represent a simple application implementation, such as installing a word processing tool on a desktop personal computer or departmental server. In contrast, Figure 2.4 is more complex and

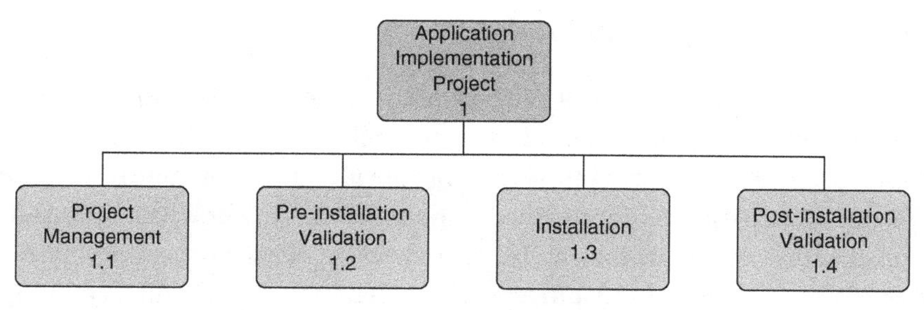

Figure 2.3 Simple WBS.

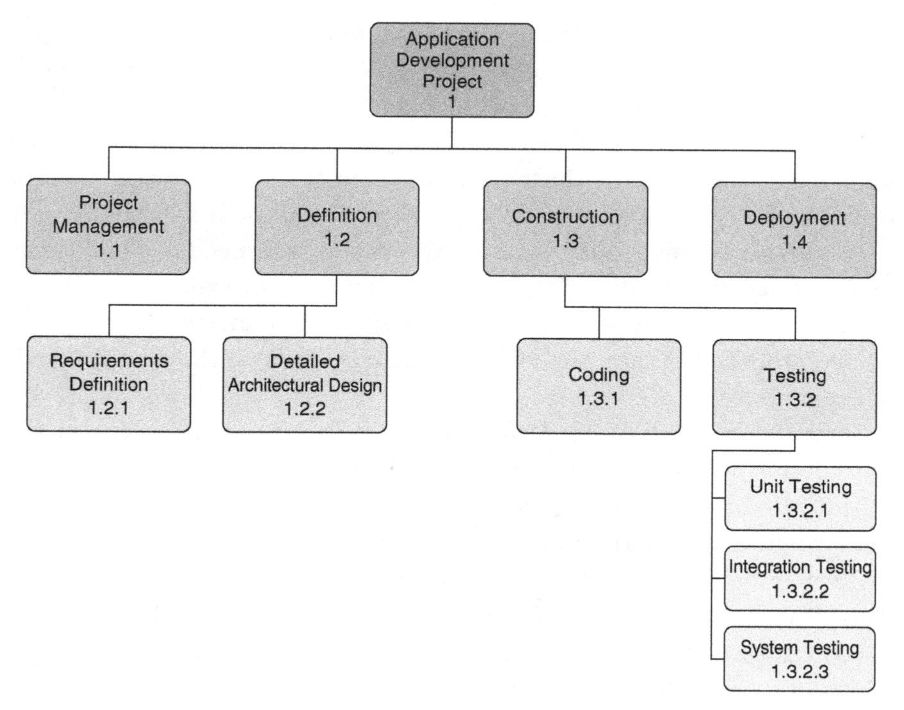

Figure 2.4 Larger WBS.

represents a custom built application, including deliverables for the definition, construction, testing and deployment of the application.

Whereas Core Characteristics apply to every project universally, Use-Related Characteristics depend on the specific requirements for each project. This implies that the quality level of the WBS correlates directly to ability of the WBS to meet the project's needs.

WBS DECOMPOSITION

WBS development can be described as a process of decomposition culminating in a level of detail that accurately captures the entire scope of the project while providing an appropriate level of detail for effective communications, management and control. But how much detail is appropriate? The real answer is, "It depends." This is the very essence of the WBS Use-Related Characteristics. The level of detail depends on the needs of the project, with the project manager working to strike the proper balance between complexity, communications, risk and the need for control.

As an illustration, let's assume we have two similar projects to implement. The first project is staffed with an experienced Project Manager and project team that has worked together previously on similar projects. The second project is staffed with an experienced Project Manager, but an inexperienced team who has never done the type of work. In this example, the Project Manager for the second team will most likely prepare a more detailed WBS as the project team will need the additional levels of detail. The WBS for the first project will probably remain at a higher level as the project team is experienced and knows what needs to be accomplished. Again, the level of detail depends on the specific needs of the project.

Interestingly, full WBS decomposition does not always occur at the start of the project. In many large, complex projects, the project team may initially decompose the WBS partially, with full decomposition occurring later when more information is known. Alternatively, some parts of the WBS may be decomposed fully while other parts are decomposed later. This "rolling wave" (*PMBOK® Guide*—Third Edition, p. 374) style of decomposition illustrates how the WBS can be utilized to meet the needs of the project, even when those needs change during the life cycle of the project.

No matter what level decomposition is right for a project, the Core WBS characteristics always apply. Chief among these is the consistent application of the 100% rule from the top through the bottom of the WBS. At all levels, the work of the child nodes must always represent 100% of the work of the parent node. In addition, it should be remembered that decomposition of elements within a WBS hierarchy need not be the same number of levels. It is perfectly acceptable to have varying levels of detail within a single WBS hierarchy.

The logic of WBS decomposition is also something that varies from project to project. The most common forms of WBS decomposition are breakdowns by the following:

- Function
- Role
- Method
- Deliverables (components)

In a functional breakdown of a WBS, the project's deliverables are grouped by business function while the deliverable orientation of the WBS is retained. This form of breakdown helps facilitate communication of responsibility to the stakeholder organizations involved in the project. Similar to functional breakdowns, role-based breakdowns also facilitate communications of responsibility for deliverables. Method-oriented groupings of the work typically organize the project's deliverables based on a defined methodology or delivery process. This, in turn, helps facilitate understanding of the project's outcomes for the project team and other project stakeholders. Breakdowns by higher-level deliverables or components are very common and used across many industries and project types. This form of decomposition is independent of the project organization or execution methodology. In many instances, this keeps the WBS simple and straightforward. In many instances, the choice of which breakdown logic to utilize is defined by the standards of the executing organization. In other instances, the Project Manager is free to choose. Whichever option is chosen, the selection of which breakdown logic to use is yet another Use-Related Characteristic for the project or program in question.

Figure 2.1 depicts a deliverable/component view of the House example WBS. Figure 2.5 shows an alternate representation of the same WBS organized by function. Both representations contain the same Work Packages, the lowest-level WBS elements. The primary difference is the organization of the higher level WBS elements.

In quality Work Breakdown Structures, each work package should represent a discreet deliverable of the project, be it product, service, result or outcome. Work Packages differ from tasks and activities in that they are deliverables expressed in the noun-object form. As part of the transition from the WBS to the Project Schedule, each work package will be further decomposed into tasks, activities and milestones which are

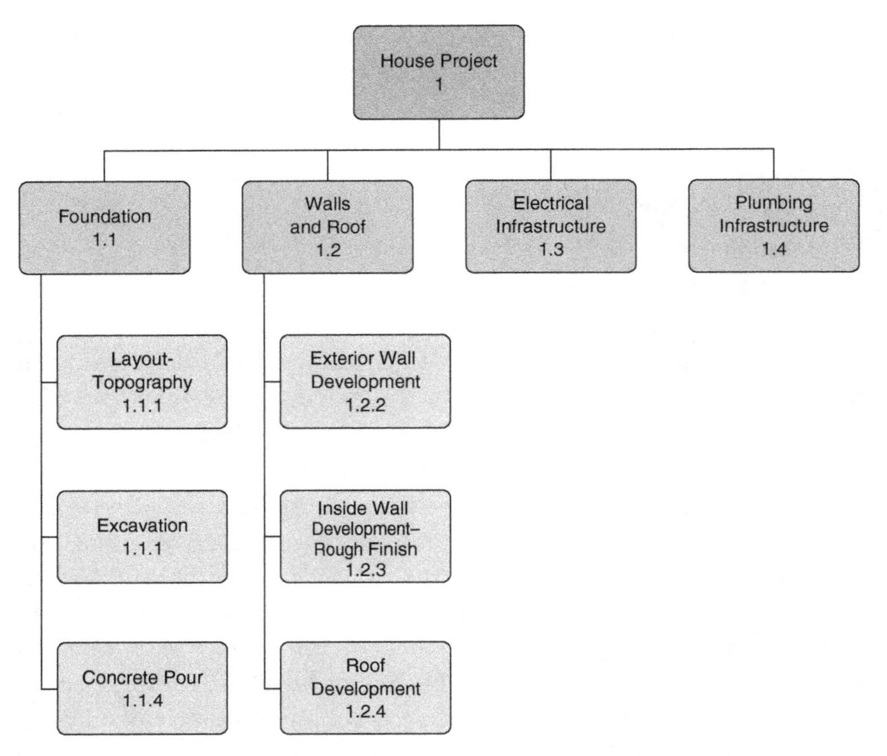

Figure 2.5 Alternative house example WBS decomposition.

expressed in the verb-object form. These tasks, activities and milestones are not part of the WBS but rather part of the Project Schedule. This is discussed in further detail in Chapter 7.

WBS IN PROJECTS, PROGRAMS, PORTFOLIOS, AND THE ENTERPRISE

Work Breakdown Structures have long been the foundation for managing individual projects. In recent years, there has been a growing trend to use the WBS to help plan and manage programs and portfolios. As WBS usage in programs and portfolios begins to take hold, we believe that WBS application will expand to the enterprise level. The PMI *Practice Standard for Work Breakdown Structures*—Second Edition recognizes this trend of WBS usage beyond the project level by defining a quality principle for it. This second WBS quality principle states that WBS quality characteristics apply at all levels of scope definition.

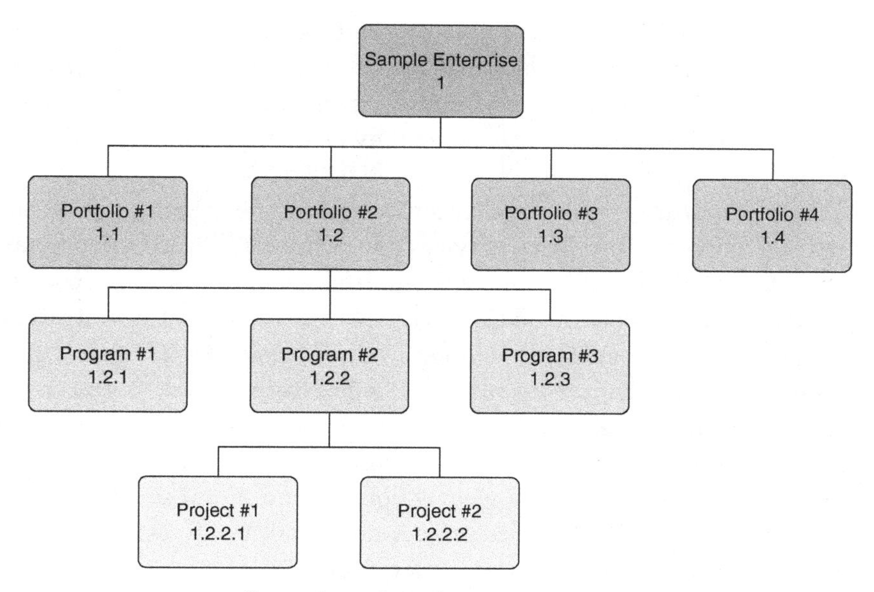

Figure 2.6 Sample enterprise WBS.

As depicted in Figure 2.6, the relationship between projects, programs, portfolios and the enterprise follow the same rules and characteristics as a single project WBS. One or more projects can be considered part of a program, which along with the other programs, are part of a portfolio. In turn one or more portfolios then make up the investments of a particular enterprise.

Earlier in the chapter we saw that there were multiple ways to decompose the WBS for a project. This is also true with programs and portfolios. Each can be organized in a myriad of ways including grouping by line of business, organization or function. The important aspect to remember is that the logic of the decomposition meets the needs of the business. Just as the work packages for an individual project roll up to the project level, so too can they roll up to define the scope of a program or portfolio. In this context, programs and portfolios are also considered higher-level deliverables that in turn are decomposed into more detailed deliverables (complete projects) further down the hierarchy. Additionally, the 100% Rule should be used to ensure that all decompositions of portfolios and programs make up 100% of the scope of the parent element.

The higher the level of the project-program-portfolio-enterprise hierarchy, the more difficult it becomes to define the complete set of deliverables. The sheer size and complexity of many organizations make this type of analysis difficult. But while this effort may be difficult, it can yield great benefits.

A highly mature, projectized organization is one where *projects* are effectively delivering products, services and results. These in turn are part of *programs* that are delivering the required benefits to the sponsoring organization. These benefits in turn are part of the organization's strategic plan and objectives which are defined by the various portfolios in which the enterprise has chosen to invest. If effectively linked, the scope of a particular project can be directly traced to the benefits delivered by the over arching program which is a component of the organization's strategic plan. This linkage, if clearly defined and maintained, can help an organization ensure that its project and program investments yield maximum return for the investors and owners.

The WBS can be an effective tool for helping organizations plan, communicate and manage the various portfolios, programs, and projects across an enterprise hierarchy. Just as there are multiple approaches for decomposing projects, so too are there many viewpoints with which enterprise- and portfolio-level Work Breakdown Structures can be organized. These viewpoints can be represented as groupings of work based on organizational, strategic, financial, cultural or risk emphasis.

WBS REPRESENTATIONS

A primary attribute of Work Breakdown Structures is that they are created to communicate scope to the various project or program stakeholders. It follows, therefore, that to communicate scope to diverse stakeholder groups, different WBS representations may be required.

There isn't a single or exclusive "correct" representation for Work Breakdown Structures; however, they are commonly represented in one of three familiar views—outline, tabular or tree structure. Each view in turn can take many forms. In Chapter 7, a representation will be introduced that provides a new approach for representing project scope. This new approach, the Scope Relationship Diagram, brings to the forefront the concept of *Inclusion* as it relates to scope. The figures in the remainder of this section will depict several common representations of the House metaphor described in Chapter 1.

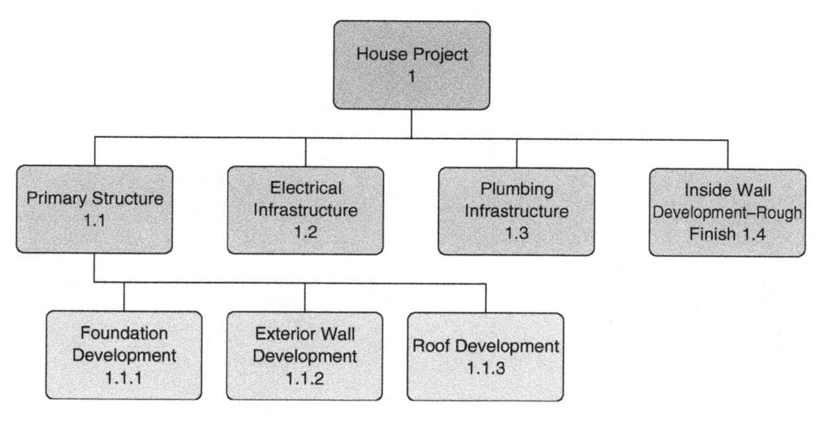

Figure 2.7 Organization chart style WBS.

The most common view of a WBS is the inverted tree structure or organization chart view as depicted in Figure 2.7. In this version of the WBS, a typical organization chart type structure is employed. Here, the root of the tree is the top and the WBS is decomposed vertically toward the bottom of the diagram.

While the typical organization chart type structure is very common, it is not the only representation of a tree structure view. In Figure 2.8, the structure is modified such that the root of the tree is at the left and the decomposition moves horizontally to the right.

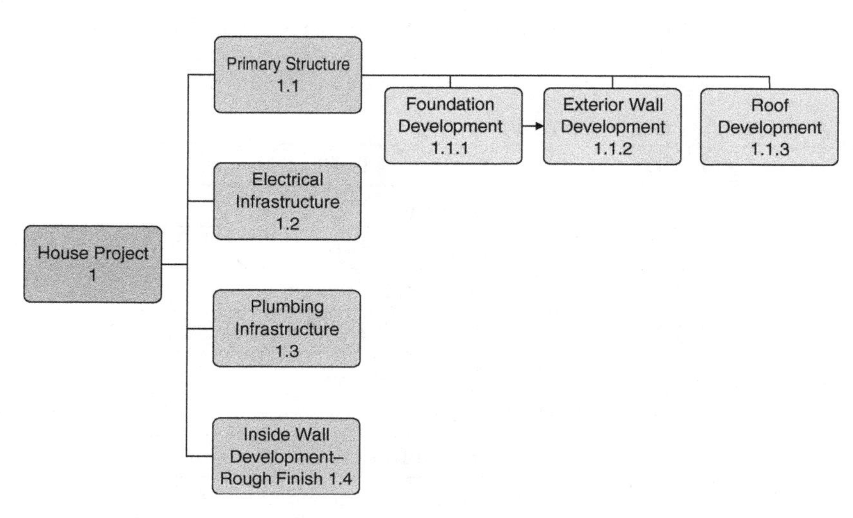

Figure 2.8 Horizontal WBS.

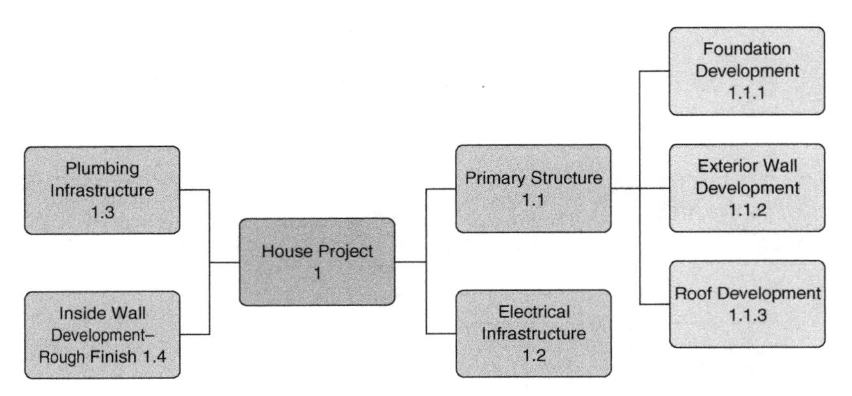

Figure 2.9 Centralized tree structure WBS.

The least common type of tree structure view is the centralized representation. This type of WBS diagram can be very useful during facilitated WBS development sessions. Figure 2.9 shows an example of a centralized tree structure view. In this example, the root of the tree is at the center of the diagram with decompositions of the WBS flowing outward from the center.

Although graphical diagrams are often used to represent work breakdown structures, they are not the only views that can and should be used. Quite often, outline or tabular views can more accurately depict the true nature of the WBS to stakeholders. Table 2.1 shows a typical outline view of a WBS.

Tables 2.2 and 2.3 show examples of tabular views of a WBS.

Table 2.1 Outline WBS View

Level	WBS Code	Element Name
1	1	House Project
2	1.1	Primary Structure
3	1.1.1	Foundation Development
3	1.1.2	Exterior Wall Development
3	1.1.3	Roof Development
2	1.2	Electrical Infrastructure
2	1.3	Plumbing Infrastructure
2	1.4	Inside Wall Development—Rough Finish

Table 2.2 Tabular WBS View #1

Level 1	Level 2	Level 3
1 House Project		
	1.1 Primary Structure	
		1.1.1 Foundation Development 1.1.2 Exterior Wall Development 1.1.3 Roof Development
	1.2 Electrical Infrastructure	
	1.3 Plumbing Infrastructure	
	1.4 Inside Wall Development—Rough Finish	

Table 2.3 Tabular WBS View #2

1 House Project
1.1 Primary Structure
1.1.1 Foundation Development 1.1.2 Exterior Wall Development 1.1.3 Roof Development
1.2 Electrical Infrastructure
1.3 Plumbing Infrastructure
1.4 Inside Wall Development—Rough Finish

All of the WBS representations in Tables 2.1, 2.2 and 2.3 depict the exact same WBS. Each version can be useful in communicating scope depending on the needs of the audience or stakeholders reviewing the information. Some people are more visual; some prefer tabular or outline views. The important thing to remember is the goal—accurately depicting and communicating project scope.

WBS TOOLS

There are many tools on the market that can help you create and depict a Work Breakdown Structure. Each tool has its own set of advantages and drawbacks. This section will describe the types of tools that can be used to create and manage a WBS. It will also describe the relative immaturity of existing tools on the market and describe the types of requirements desired for WBS tools moving forward. This section will not discuss nor review specific tools from vendors. The purpose of section is to provide insight into the types of tools that can be employed for WBS creation and management.

There many types of tools in the market today, both low and high tech, that can be employed to create and depict a Work Breakdown Structure. One of the original low-tech tools used in the planning process was simple paper and pen. In more recent decades the paper and pen have given way to colored sticky notes and white board markers. Colored sticky notes, while very low-tech, are still one of the most effective methods for group facilitation and creation of a WBS. Colored sticky notes can be used to capture individual project deliverables. With the use of an open wall or white board, they can then be logically grouped to help visually create a hierarchical WBS.

With the advent of personal computers and laptops, high-tech tools have now become the most common method for creating and depicting Work Breakdown Structures. The table in Table 2.4 shows the most common types of information technology tools used to create WBSs and the relative advantages and challenges of each.

The rapid growth of the project scheduling tool market saw a major increase in the use of project scheduling tools to create Work Breakdown Structures. Although project scheduling tools, as they exist today, provide for integration between the WBS and the Project Schedule, all too often they can lead to problems. Using a project scheduling tool can make it very difficult to differentiate between WBS elements (deliverables) and project schedule tasks, activities and milestones. The concept that Work Packages are decomposed into tasks, activities and milestones during the transition process from WBS to Project Schedule is something that is lost in the current set of tools. In addition, when using project scheduling tools to depict the WBS, all too often project managers create task lists that do not accurately represent all of the work and are task (action) oriented, rather than deliverable-oriented.

Table 2.4 WBS Technology

Type of Tool	Advantages	Challenges
Project Management Scheduler	• Integration of the WBS with the Project Schedule	• Failure to create a high-quality, deliverable oriented WBS in favor of a task-oriented one • Difficult to differentiate between WBS elements and Project Schedule elements • Inability to accurately bound and report on scope elements due to tight integration with Project Schedule
Spreadsheet	• Excellent tool for creating tabular WBS views • Ability to create and manage large, complex WBSs • Ability to integrate WBS with WBS Dictionary	• Lack of a visual representation of the WBS
Word Processor	• Good tool for integrating multiple WBS views, including a WBS Dictionary	• Difficult to scale for large, complex WBSs
Graphics Development	• Good tool for creating visual WBS views	• Difficult to scale for large, complex WBSs
Enterprise Project Management (EPM)	• Integrates many aspects of project management (scope, scheduling, cost, etc.)	• Large, complex, and difficult to implement • Limited application for small or intermediate projects • Costly • Still evolving; not as sophisticated as they might be

With the growing popularity of office application suites, spreadsheets, word processors and graphics development applications, many of these are increasingly being used to create varied and customized WBS views. On large projects, spreadsheets can be used to capture the full size and complexity of the WBS. Additionally, spreadsheets provide for the integration of the WBS and WBS Dictionary. As a result, spreadsheets are being used to create Work Breakdown Structures and WBS Dictionaries more frequently. With graphics development tools becoming easier to use and visual representations becoming more popular, other technologies such as word processors, databases and reporting tools can also be used for creating, managing and representing Work Breakdown Structures.

Even with all of the technology that exists today, WBS tools are still relatively immature. The authors of this book would like to encourage vendors to focus on creating tools that provide the ability to adequately capture and manage WBS elements, highlighting and focusing on the ability to differentiate WBS elements from Project Schedule elements. Work Breakdown Structures are the very heart of project and program management. Tool vendors, and ultimately every industry, may be better served when tools exist that allow project and program managers to properly capture, manage and report on scope, while enabling integration with other project management scheduling tools.

CHAPTER SUMMARY

This chapter detailed many important concepts regarding the WBS. The opening sections discussed the attributes and characteristics that produce quality Work Breakdown Structures. Core characteristics define the minimum set required in all Work Breakdown Structures while Use-Related Characteristics bring to bear those that differentiate one project from the next.

The central sections discuss WBS decomposition and the use of Work Breakdown Structures at all levels of scope definition and management. We note how the attributes and characteristics that make up a quality WBS apply at all levels of scope, whether they are applied to project, program, portfolio or the enterprise. The section on WBS representations dispels the myth that a WBS must be represented by an inverted tree, organization-chart-style representation. Several different representations are provided, including graphical, textual and outline views. The chapter concludes with a discussion of tools to aid in WBS creation and representation. In this section, we note that with all the technology that exists today, WBS tools remain relatively immature. The authors encourage vendors to continue to accelerate the evolution of WBS tools, such that they adequately capture and manage WBS elements, separate and distinct from Project Schedule tasks and activities.

REFERENCES

Haugan, Gregory T. 2002. *Effective Work Breakdown Structures*. Vienna, VA Management Concepts.

Pritchard, Carl L. 1998. *How to Build a Work Breakdown Structure: The Cornerstone of Project Management*. Arlington VA. ESI International.

Project Management Institute. 2004. *A Guide to the Project Management Body of Knowledge (PMBOK® Guide—Third Edition)*. Newtown Square, PA: Project Management Institute.

Project Management Institute. 2004. *Errata to the Project Management Body of Knowledge (PMBOK® Guide—Third Edition)*. Newtown Square, PA: Project Management Institute.

Project Management Institute. 2006. *Practice Standard for Work Breakdown Structures—Second Edition*. Newtown Square, PA: Project Management Institute.

CHAPTER QUESTIONS

1. Which of the following are Core Characteristics of a quality WBS? (Select all that apply.)
   a. Deliverable oriented
   b. Task oriented
   c. Hierarchical
   d. Includes only the end products, services, or results of the project
   e. Uses nouns, verbs, and adjectives
   f. Is created by those performing the work
2. Which of the following is true for quality WBS?
   a. Program/Project Management can occur at any level of the WBS.
   b. It contains at least three levels of decomposition.
   c. It clearly communicates project scope to all stakeholders.
   d. It does not include a WBS Dictionary.
3. Which of the following is true for WBS Use-Related Characteristics?
   a. Characteristics are consistent from project to project.
   b. WBS quality depends on how well the specific content and types of elements address the full set of needs of the project.
   c. It contain only discrete WBS elements.
   d. It must be decomposed at least three levels.
4. Which of the following statements is true for any WBS? (Select all that apply.)
   a. WBS quality characteristics apply at all levels of scope definition.
   b. Valid WBS representations include only graphical and outline views.
   c. Using a project scheduling tool for WBS creation is helpful in differentiating between WBS elements and Project Schedule tasks and activities.
   d. Tools for creating and managing WBS are very mature and easy to use.

Part II

WBS Application In Projects

Chapter 3

Project Initiation and the WBS

CHAPTER OVERVIEW

Initiating, as defined in the dictionary, means "to begin, set going, or originate" (http://www.dictionary.com) and this chapter is just that. Here we will discuss the beginning or start of a project and how the WBS interacts in and plays a part in that origin. In addition, we will show how, from the beginning, a Project Manager (or other interested parties) can trace a project's scope throughout the project life cycle and artifacts. Specifically, this chapter will discuss these topics:

- Project Charter
- Preliminary Project Scope Statement
- Contracts, Agreements and Statements of Work (SOW)

All of these items will build on each other and provide a lead in for the next chapter, where we move into Planning. In order to guide you through the visualization of increasing levels of detail, traceability and elaboration of project scope from artifact to artifact, we have added a triangle icon in this chapter and in Chapter 4.

The first appearance of the icon is the completed version. As we begin this chapter, you will notice the icon is mostly blank. We will fill and complete the icon as we move from concept to concept and from this chapter into Chapter 4. So, let us begin.

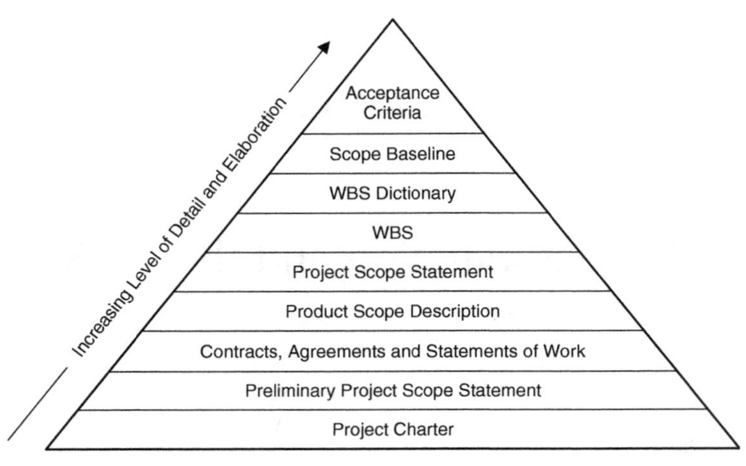

Project Scope Definition and Elaboration

PROJECT CHARTER

If you look up the definition of a charter, it can be traced back hundreds of years to when it was understood to be the following:

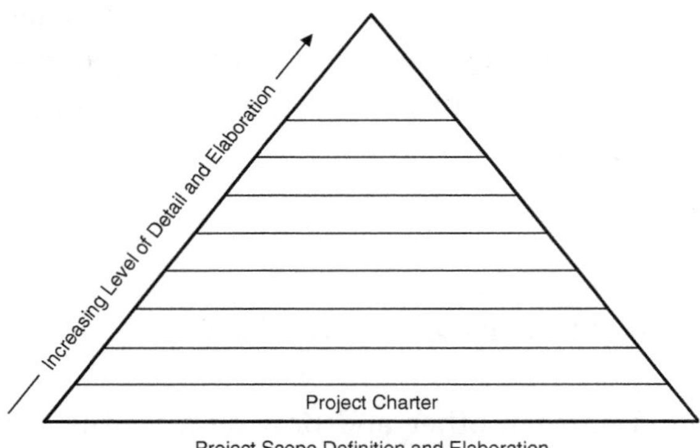

Project Scope Definition and Elaboration

> A document granting certain rights, powers, or functions. It may be issued by the sovereign body of a state to a local governing body, university, or other corporation or by the constituted authority of a society or order to a local unit. The term was widely applied to various royal grants of rights in the Middle Ages and in early modern times. The most famous political charter is the Magna Carta of England. Chartered companies held broad powers of trade and government by royal charter. In colonial America, chartered colonies were in theory, and to an extent in fact, less subject to royal interference than were royal colonies.
>
> *(Columbia Electronic Encyclopedia 2008).*

Today, the definition has changed only slightly in that it is still considered a document, issued by a governing body (many times a project sponsor or company executive) that formally authorizes work to be completed. The charter, by today's terms, continues to outline what is expected of the body to which it is granted and provides a designated person (usually the Project Manager) the authority to complete the work enclosed in the chartering document.

All of that may sound pretty heavy, but the point is that the concept of a charter is not new, just the use has changed slightly in the last hundred

Project Overview

This project is being undertaken to establish a new residence for Mr. and Mrs. John Smith. The new residence will be a free-standing, single-family dwelling built on a two-acre lot (lot #24) located at 200 North Maple Avenue, MyTown, MyState, 20001-1234, USA. The project is to commence on Monday, February 2, 2015 and will complete no later than Thursday, December 31, 2015.

This home is being constructed to take advantage of the latest building materials and codes and will employ emerging technology to minimize energy consumption. Construction will be overseen and managed by Apex Home Builders, the prime contractor who may subcontract components of the construction effort.

All labor will be bonded and all materials will meet or exceed local building code guidelines.

Section I. Project Purpose

The home project is being undertaken to establish a new primary residence for Mr. and Mrs. Smith and family. The new residence is scheduled for completion in December so that the Smith family may move in during the first two weeks of 2016. Mr. Smith will be taking responsibility for his company's North American operations in 2016 and is relocating from Europe to do so. Mr. Smith and family will be traveling and relocating during December of 2015 and will move directly from their current home to the newly completed residence.

The home must be completed by December 31, 2015 so that the Smith's can establish residence in the community with the appropriate lead time to enable their children to be enrolled in the school system to begin the 2016 school year along with their classmates.

Exhibit 3.1 House Example Project Charter

or so years. Today, the **Project Charter** still defines the high-level scope of work to be completed, documents why or the need for the completion of the project and defines the products, services or outcomes to be delivered to the customer.

So you may be asking yourself (or us)—how does the WBS fit into this? The answer to that question is that the WBS is used as a tool for scope definition, and without the Project Charter providing the boundaries for the scope, it would be nearly impossible to create a valid WBS. This is due to the fact that in order for the WBS to "organize and define the total scope of the project" (definition from the *PMBOK® Guide*—Third Edition), it must have the scope defined in the Project Charter. It is, therefore, very important for the team creating the WBS to have an approved Project Charter available to work from and to provide direction. Without it, the team could easily be moving in any direction, and perhaps not the right direction. This is also the beginning of the project scope traceability path. The following exhibit depicts the beginnings of a Project Charter for the House example. A full copy of this sample Project Charter can be found in Appendix A at the end of this book.

Before we can get to the WBS, there is still a bit of initiating work that must be completed in order to have an accurate picture of the project's scope before that scope can be fully detailed and articulated. The next step in our journey is the elaboration of the Project Charter into the development of the Preliminary Project Scope Statement.

PRELIMINARY PROJECT SCOPE STATEMENT

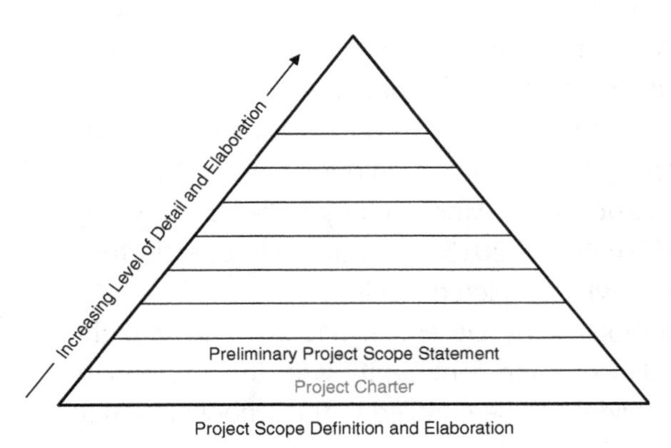

Project Scope Definition and Elaboration

The **Preliminary Project Scope Statement** will be used to set the context for the planning work to be accomplished in our next chapter. The Preliminary Project Scope Statement should be used to further elaborate

Project Scope Statement

Project Overview

This project is being undertaken to establish a new residence for Mr. and Mrs. John Smith. The new residence will be a free-standing, single-family dwelling built on a two-acre lot (lot 24) located at 200 North Maple Avenue, MyTown, MyState, 20001-1234, USA. The project is to commence on Monday, February 2, 2015 and will complete no later than Thursday, December 31, 2015.

This home is being constructed to take advantage of the latest building materials and codes and will employ emerging technology to minimize energy consumption. Construction will be overseen and managed by Apex Home Builders, the prime contractor who may subcontract components of the construction effort.

All labor will be bonded and all materials will meet or exceed local building code guidelines.

Section I. Project Purpose

The home project is being undertaken to establish a new primary residence for Mr. and Mrs. Smith and family. The new residence is scheduled for completion in December so that the Smith family may move in during the first two weeks of 2016. Mr. Smith will be taking responsibility for his company's North American operations in 2016 and is relocating from Europe to do so. Mr. Smith and family will be traveling and relocating during December of 2015 and will move directly from their current home to the newly completed residence.

The home must be completed by December 31, 2015, so that the Smith's can establish residence in the community with the appropriate lead time to enable their children to be enrolled in the school system to begin the 2016 school year along with their class mates.

Section II. Project Scope

This is a Fixed-Price Contract

Contractor commitment estimate is U.S.$750,000.00.

(continues)

Exhibit 3.2 House Example Preliminary Project Scope Statement

(continued)

Upon completion, the new property will include the following as described in the detailed specifications and blueprint:

- Landscaping
- Foundation (with basement)—poured concrete and concrete block
- Driveway—2000 feet, concrete with brick inlay
- Main home—4500 square feet, brick/stucco
- Deck / Patio / Screen Room
- Garage—1600 square feet, two story

Section III: Project Milestones

As described in Section II, completion of the project must be achieved by December 31, 2015. Progress milestones associated with the project are as follows:

1 Architectural drawings complete and approved
2 Building permit approved
3 Lot preparation and clearing complete
4 Foundation excavation complete
5 Footings poured and set
6 Foundation poured, block construction complete, foundation set
7 Home and garage exterior closed to weather
8 Driveway and landscape complete
9 Interior wiring complete
10 Exterior wiring complete
11 HVAC complete
12 Interior plumbing complete
13 Exterior plumbing complete
14 Interior finish complete
15 Exterior finish complete
16 Walkthrough complete
17 Certificate of Occupancy granted
18 Interior and exterior punch list approved
19 Interior and exterior punch list complete
20 Acceptance review and key turnover complete

and define the information provided in the Project Charter. It also documents the characteristics and boundaries of the project in more detail, answering the question, "What dowe need to accomplish?" Other items covered in the Preliminary Project Scope Statement include how the final product, service or outcome will be *measured* in order to gain acceptance of the final product, how scope will be controlled, how the initial project organization will be formed, the initial risks identified as part of the project and ultimately, an order of magnitude cost estimate. The Preliminary Project Scope Statement is another input into the creation of the WBS and it will be further refined during Scope Definition. The following exhibit depicts the beginnings of a Preliminary Project Scope Statement for the House example. A full copy of this sample Project Scope Statement can be found in Appendix B at the end of this book.

CONTRACTS, AGREEMENTS, STATEMENTS OF WORK (SOW)

Additional inputs to the creation of the WBS are contracts, agreements and **Statements of Work**. Each of these should be handled in similar ways. Contracts define the financial and legal accountabilities between buyer and seller, project team and customer or perhaps a project sponsor and project team. Although it is not an ideal situation to have a contract in place before the project is fully defined, it does happen and should be used as an input into the scope definition processes as this work must be completed during the execution of the project.

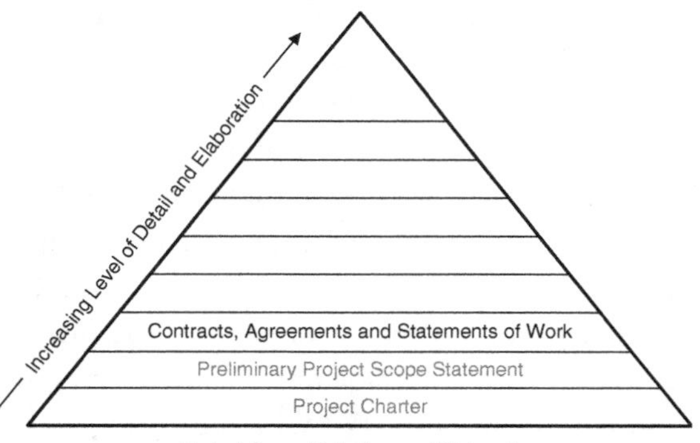

Project Scope Definition and Elaboration

In contrast to the nature of contracts, agreements are much less formal. Agreements are typically informal arrangements between two parties regarding a specific course of action. While they may not be as formal, agreements are considered to be legally binding. A project team should not overlook informal agreements as inputs to the WBS as they are also a method of documenting the scope of the project.

A Statement of Work (SOW) is "a narrative description of products, services, or results to be supplied" (*PMBOK® Guide*—Third Edition, p.376) by the project and project team. The SOW should be derived from the Preliminary Project Scope Statement. By the same token, the Contract Statement of Work is defined as "a narrative description of products, services, or results to be supplied under contract" (*PMBOK® Guide*—Third Edition, p. 355). This Contract Statement of Work is utilized to define the scope of work to be completed by subcontractors.

Both the Statement and Contract Statement of Work are derived from the Project Charter and the Preliminary Project Scope Statement. They should include detailed descriptions of what the project will produce as deliverables and should include the business need, product and service attributes and a description of the project scope. The Contract Statement of Work may also include a definition of other project-process deliverables, such as performance reporting and post project support.

As each of these project artifacts is completed in progression, they further define the original project scope set out in the Project Charter, building to the WBS, which will become, along with the Scope Statement a part of the scope baseline.

The Project Charter, the Preliminary Project Scope Statement, contracts, agreements, the Statement of Work and the Contract Statement of Work all become inputs to the work to be completed in the next chapter—that is Scope Definition and the work to complete the Work Breakdown Structure and WBS Dictionary.

Each of these items is important to initiating projects properly. Without this foundational information, all else that follows could be considered less than complete. We will continue the development of our project in the next chapter—using each of the items that have been created here during *Project Initiation*—as we start to plan.

CHAPTER SUMMARY

A fully defined and elaborated scope definition is the cornerstone of any successful project. This is very similar to the foundation being essential to the success of our house example. By fully utilizing the foundational tools discussed in this chapter and tracing the project scope from a somewhat vague statement in the Project Charter through contracts, agreements and Statements of Work, we begin to strengthen this foundation.

In the next chapter, we will continue to elaborate on the foundation we have begun, further defining and detailing the project scope. This will culminate in a Work Breakdown Structure, WBS Dictionary and Scope Baseline.

REFERENCES

Definition of "Initiating". Available at http://www.dictionary.com.

Definition of "Charter". Columbia Electronic Encyclopedia. Retrieved January 29, 2008, from http://www.reference.com/browse/columbia/charter.

Project Management Institute. (2004). A Guide to the Project Management Body of Knowledge (*PMBOK® Guide*—Third Edition). Newtown Square, PA: Project Management Institute.

CHAPTER QUESTIONS

1. Which of the following key project documents are created in the *initiating* phase of a project?
   (Select all that apply.)
   a. Project Charter
   b. Preliminary Project Scope Statement
   c. Product Scope Description
   d. Work Breakdown Structure

2. Which foundational project management document provides the initial boundaries for the project's scope?
   a. Project Charter
   b. Preliminary Project Scope Statement
   c. Product Scope Description
   d. Work Breakdown Structure

3. Which foundational project management document is used to set the context for much of the planning phase of the project?
   a. Project Charter
   b. Preliminary Project Scope Statement
   c. Product Scope Description
   d. Work Breakdown Structure

4. Which foundational project management document includes information on how the final products, services, or outcomes of the project will be measured?
   a. Project Charter
   b. Preliminary Project Scope Statement
   c. Product Scope Description
   d. Work Breakdown Structure

5. Contracts should always be put in place before the project is fully defined.
   a. True
   b. False

Chapter 4

Defining Scope through the WBS

CHAPTER OVERVIEW

Now that the initiation of the project has been completed, the project team should have a strong understanding of the direction they have been challenged to pursue. It is now time for project planning to begin.

In this chapter, we will discuss these topics:

- Product Scope Description
- Project Scope Statement
- Work Breakdown Structure (WBS)
- WBS Dictionary
- Deliverable-Based Management
- Activity-Based Management
- Scope Baseline
- Acceptance Criteria

This chapter will provide a solid foundation for monitoring and controlling the remainder of the project. You may ask, what is there left to do? Well, let's get started and we'll explain.

PRODUCT SCOPE DESCRIPTION

First is the **product scope description**. You may be thinking that we've already discussed this step, but in actuality, we haven't. We began the previous chapter with a Preliminary Project Scope Statement, which

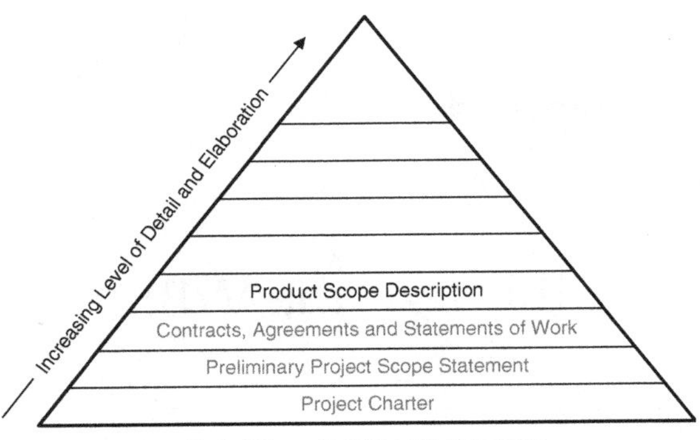

Project Scope Definition and Elaboration

we are about to refine. That Preliminary Project Scope Statement covers the work defined in the Project Scope "that must be performed to deliver a product, service or result with the specified features and functions" (*PMBOK® Guide*—Third Edition, p. 370). The last part of that definition, the "specified features and functions," is the *product* scope definition. This particular part of the Scope Statement will provide and describe the unique characteristics of the *product* that is being produced as a result of the project defined in the charter. The product scope description provides, in narrative format, the *look and feel* of what is being produced as well as the benefit the program is delivering. This will then be utilized to ensure that the work required to complete the product described is included in the WBS. The product scope description should be developed in parallel with the project scope statement. If there are changes to scope related to changes in the product's features and functions, they should be reflected in the product scope description as well as in and aligned with the project scope statement and project acceptance criteria.

PROJECT SCOPE STATEMENT (SCOPE DEFINITION)

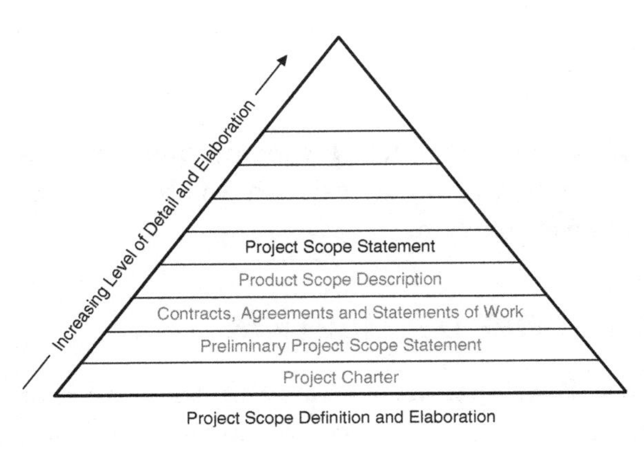

Project Scope Definition and Elaboration

At the same time the Product Scope definition is being completed, the Preliminary Scope Statement should be revised in order to describe what is to be delivered by the project as well as the work required to complete those deliverables. These two documents will become the basis for more detailed planning starting with the creation of the WBS. The WBS will be used as a guide and basis for monitoring and controlling the project throughout execution and closing. The **Project Scope Statement**, the Product Scope Description and WBS will become the baseline for evaluating changes to the project (as described in Chapter 8).

WORK BREAKDOWN STRUCTURE

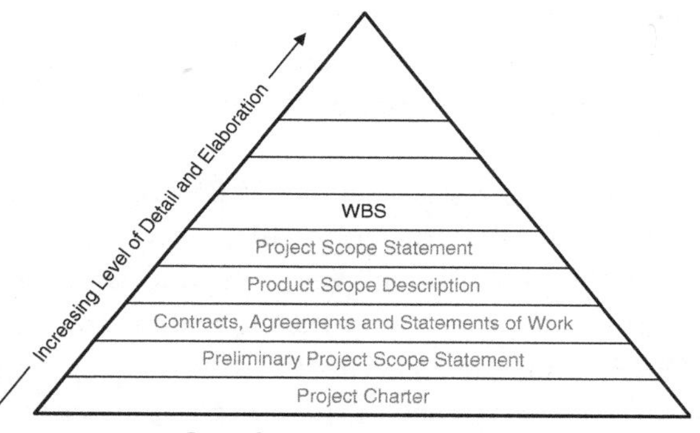

Project Scope Definition and Elaboration

The key to project success is to ensure all project stakeholders and team members have a common understanding of what the project is to achieve. In order to do that, the following questions must be asked and answered–why, who, what, when, where and how? While that doesn't follow the familiar cadence of how those questions are typically asked, we believe we should answer the *why* question first–and since we're the authors, we can do that.

So, *why* perform the project and moreover, *why* create the WBS? A common claim we hear from project managers is "I have a lot of work to get done in a very short period of time. Why would I take the time to create a WBS? I know what I have to do–just get it done–right?" Well, maybe. If you are not concerned about the result of the project, this is the correct answer. In reality, it is important to take the time to complete the WBS because doing so will help the Project Manager determine and detail everything that must be completed in order to achieve and deliver the project objectives.

For example, envision a project you have worked on recently, one in which a WBS was not completed. Think about what was known at the beginning of the project and then reflect on what was delivered at the end. Were they the same?

Now, envision that same project with a WBS created as discussed in Chapter 2, WBS Core Characteristics. This WBS would allow the project manager and *those involved in the project* to create a *deliverable oriented*, *hierarchical* representation of the entire project scope. It would include all of the work to be delivered by the project, *all project-related work elements including all internal, external and interim deliverables*. How much more successful would the original project have been using the latter scenario? It is truly difficult to predict, but the authors suggest the outcome would be substantially more controlled, if not improved. The WBS in this instance can be traced directly to the project scope and if utilized correctly, to all succeeding artifacts in the project life cycle.

This also answers the question of *what* should be included in the WBS. The WBS should include all the work that will produce the intended outcomes and nothing but that work. All deliverables identified in the WBS will be produced by the project and accordingly, work excluded from the WBS will not.

Without a well-developed WBS created at the project start and updated throughout the project, the likely success of any project can be diminished. Now that we have answered the question about why the WBS is essential to a given project, the next question to address is *who?* Who owns the WBS, who should be involved in developing it, who keeps it up to date? The answer to each of these is the project manager and the project team. As one of the Core Quality Characteristics, the WBS "is created by those performing the work with technical input from knowledgeable subject matter experts and other project stakeholders." (*Practice Standard Work Breakdown Structures*—Second Edition, p. 20) This is absolutely critical to the success of any WBS. So, who is responsible for the upkeep of the WBS? This is the responsibility of the entire project team. The WBS should be created early in the life cycle of the project and should mature as more becomes known about the project scope and throughout the execution of the project. This will be covered in detail in later chapters. One thing to take away from this section–the WBS is a living document and should reflect the project scope throughout the project life cycle.

So, *when* should the WBS be created–and how often is it updated (*where*)? Early and *continuously*! If the WBS is created early, it allows for a conceptual view of the project by all project participants and stakeholders. As the project definition matures–through the exercises and documentation discussed in Chapter 3, the WBS will be updated continually to the point of scope baseline. Once baselined, the WBS is placed under change control along with the project scope documents. This produces a controlled environment (baseline) for the project and the Project Manager from which he or she can closely manage potential changes to the *project* or *product* scope. In addition, this allows for the generation of the subsequent project components, such as the Project Schedule, the assignment of work and resources, communications plans, change management, financial management, scope management and risk and issue management. As each of these items are employed as part of the project, the WBS is updated to reflect changes to the evolving scope, allowing for a stable environment with controlled change for the development of the project deliverables.

So we come to the last question–*how?* How do you create a Work Breakdown Structure? To begin, the Project Manager and team must clearly understand what the scope defines. And as we have noted earlier, a solid starting point are the artifacts created in Chapter 3 of this book–the Project Charter, Preliminary Project Scope Statement, contracts, agreements and Statement(s) of Work. By reviewing and analyzing these documents, the team can begin to understand the direction and boundaries of the project and can initiate the processes necessary to define and detail the work and deliverables that will be completed in order to deliver the desired outcomes.

Next, the project manager must determine what other tools and resources he or she has available for use. Is there a library of templates within the company? Have previous projects employed a particular WBS template that might fit appropriately as a starting point for this project? Has the project manager used another tool or template from a previous project that would be an enabler for success? Are templates from colleagues available? Each of these may be a valuable resource that can provide a strong starting point and can help with the development of the WBS.

Finally, the team needs to determine how it will build the WBS. There are four traditional approaches to creating the WBS (Table 5.1, WBS Creation Methods, *Practice Standard for Work Breakdown Structures*—Second Edition, p. 29). These include Top-Down, Bottom-Up, Standards and Templates. We will discuss each of these, starting with Standards and Templates. For companies and organizations where they exist, standards and templates provide a common format and preferred methodology for developing each project's WBS. These items are predefined and may be required by the organization or company. While deviating from the standard may not be an option, using a standard or template can be very helpful to someone new to Project Management and/or unfamiliar with the development and use of Work Breakdown Structures. Using existing standards and templates also enhances and ensures consistency across projects within an organization. While these tools often provide consistency and ease the development of project artifacts for some project managers, others may find the use of existing standards and templates confining, limiting the project manager's creativity. Project Managers who have "been there, done that" may feel that using the existing tools does not allow the flexibility or creativity that some desire in creating a WBS. Additionally, employing templates or standards may appear to the project manager as forcing a "square peg into a round hole" because the existing standards or templates may not easily fit the type of project or project structure. When this is the case, slight modifications to the templates and standard protocol may be necessary to meet the needs of the project.

By utilizing—or, more importantly, by exclusively relying on existing templates or standards for the creation of a WBS, it is possible to inadvertently include unnecessary deliverables or work outlined by the WBS template that is truly not a part of the project. The opposite may also be true. The template or standard may exclude important work that must be performed to reach the desired outcomes. When these oversights are not addressed, the WBS will not provide an effective foundation for the project deliverable(s) it is designed to produce. One way to ensure this does not occur is to conduct multiple reviews of the WBS. Preferably these reviews would be performed by a variety of project personnel. Though it is possible to allow for a single review by one project team member, we recommend a group read-through of the WBS, line by line, to make 100% sure each individual WBS component is appropriate for inclusion in the WBS and can be traced back to the Project Charter, the Preliminary Project Scope Statement, the applicable contracts, agreements, Statement(s) of Work, the Product Scope Description and the Project Scope Statement. WBS elements should be added, updated or deleted as appropriate based on this review cycle.

Other methods of creating the WBS include both top-down and bottom-up methods. The decision to use one or the other of these methods is usually the preference of the team creating the WBS. With top-down creation, the team starts at the top of the WBS, describing the final product(s) to be realized as a result of the entire project. From there, the team defines and decomposes the major deliverables that will be produced in order to achieve the desired product(s), service(s), or result(s). Once the major deliverables have been defined, they can be further decomposed into the work packages needed to realize each major deliverable. This decomposition process continues until a level appropriate for management and tracking of the project is achieved. The level of decomposition can vary significantly depending on many factors including the maturity of the project organization responsible for producing the final deliverables. The process of decomposition is typically iterative. To achieve the appropriate level of decomposition, the Project Manager and team should repeatedly ask "Have we included everything; and have we included anything that isn't truly defined by the project and product scope statements?" By doing this, the team will ensure that all work to be completed for the project is identified and accounted for.

In contrast, the bottom-up method of creating a WBS starts with the lowest level of deliverables and brings them together, consolidating all of the work toward the top of the WBS, to represent the final product(s) to be achieved. Using this approach, the Project Manager and team must define all deliverables and work packages before they can be logically grouped into parent-child relationships during the creation of the WBS. By developing the WBS in this bottom-up manner, the Project Manager ensures that all work packages are included. A caution here for those of you considering this approach: It is easy to become mired in the detail of defining all of the project's work packages and lose sight of the bigger picture of what ultimately needs to be accomplished to produce the final product.

No matter the method chosen, the Core and Use-Related Characteristics (defined in Chapter 2) should be followed. In addition, we would like to reiterate and reinforce the fact that regardless of the method of WBS composition, the process is iterative and is repeated throughout the initiating and planning phases of the project as more and more is information is revealed or becomes available. Many times, the project's WBS will be initially decomposed to a satisfactory level of detail for the Project Manager and Team–but as questions are answered about the scope of the project–the Project Manager and team will realize the WBS can be decomposed further. By remaining alert to this process, they can better develop and maintain the WBS throughout the remainder of the project life cycle. For an example of the WBS created for the House Metaphor, see Exhibit 4.2 (see page 62).

BEGINNING WITH THE ELABORATED WBS

We have discussed the creation of the WBS using quality principles and we have referenced writings from key authors and drawn upon PMI's *Practice Standard for Work Breakdown Structures*—Second Edition. We have reviewed tools and techniques that guide the development of Work Breakdown Structures that reflect Core Characteristics. Now we would like to discuss the application of those characteristics, so the WBS fully serves the project team throughout the delivery of projects.

Some of the most challenging aspects of Project Management include knowing when to apply a particular principle, tool or practice–and having made that determination, knowing how much or how little of the principle or tool to apply. This problem applies across the project management spectrum and touches process groups, knowledge areas, competencies and skills. It naturally follows that this issue applies to the development of foundational elements, including the Work Breakdown Structure and presents itself often as a question about the depth or detail of the WBS.

There are differing views about the degree to which a given WBS should be decomposed or elaborated. As the use and application of the WBS has expanded, there has been quite a bit of controversy about how far each leg of the WBS should be elaborated or broken down. Some practitioners and knowledge experts contend that all legs should be decomposed to the same level in every depiction, while others strongly assert that it isn't necessary to do this. Your authors fall into this latter category. In fact, we recommend that you decompose only those WBS elements that are actively in play within a project at any particular point in time. As WBS elements are decomposed during the execution of the project, supporting plans such as the Risk Management Plan, the Staffing Plan, the Project Schedule and Communications Plan should be updated accordingly. By the time the project has closed, all WBS elements should be decomposed.

Why do we recommend this? There are a number of reasons for taking a more flexible approach to the decomposition of the WBS. In order for the WBS to function as an active component of the project, it must express and detail the scope elements and deliverables all project stakeholders approve for delivery. While this is the case, some of the deliverables that would be delivered late in the project might not be fully known during the early phases–and even if they were, detailing and displaying them at all times has a tendency to clutter and confuse discussions that can easily be focused on smaller or narrowly defined work elements.

1 Project Name
  1.1 WBS Element 1
  1.2 WBS Element 2

Exhibit 4.1 WBS Outline View

The *Practice Standard for Work Breakdown Structures*—Second Edition (pg. 20) points out that the WBS should be represented by two levels of decomposition—at a minimum. That means there must be a WBS element that represents the entire project at level 1—and at least two elements at the next level of decomposition, level 2. Why do we say two elements at the next level of decomposition? The reason goes back the very essence of the WBS—the 100% Rule. With this rule, the children of a particular parent element must represent 100% of the scope of the parent element. If a particular parent element only has one child, then for the 100% Rule to remain true the parent and the child elements must in fact be duplicates of one another. If this is the case, the redundancy will serve no useful purpose and as such should be eliminated. This is why decomposing an element will always require at least two child elements. To illustrate this WBS Core Characteristic, the work should be decomposed to represent something that looks like the elements depicted in Exhibit 4.1 and Figure 4.1.

As we know, these two examples show the minimum breakdown or decomposition of the WBS–though we can imagine that if your WBS required no more elaboration than this, the usefulness of the WBS would likely be limited to simply showing how the three elements relate to one another, because the work defined by this very basic "project' is very likely

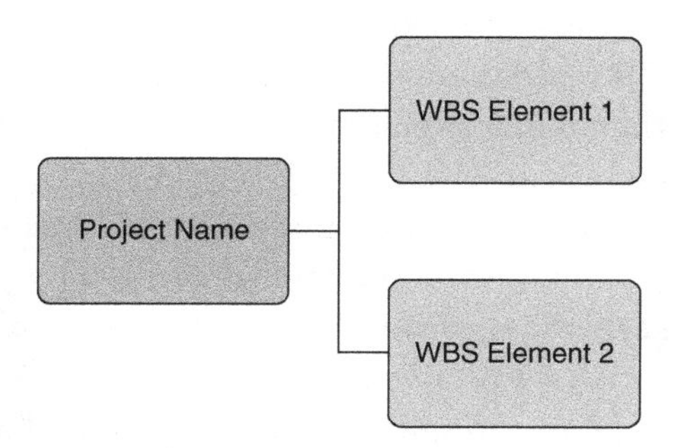

Figure 4.1 WBS Tree Structure View.

1 House Project
- 1.1 Primary Structure
  - 1.1.1 Foundation Development
    - 1.1.1.1 Layout–Topography
    - 1.1.1.2 Excavation
    - 1.1.1.3 Concrete Pour
  - 1.1.2 Exterior Wall Development
  - 1.1.3 Roof Development
- 1.2 Electrical Infrastructure
- 1.3 Plumbing Infrastructure
- 1.4 Inside Wall Development: Rough Finish

Exhibit 4.2 House Example WBS

known to all involved. And if that is the case, you can stop there–you've developed a WBS for the project that illustrates the breakdown of all the work as well as the relationship between the work elements listed.

USE-RELATED CHARACTERISTICS

Because a majority of projects are considerably more complex than the examples in Exhibit 4.1 and Figure 4.1, the level of detail shown will not be enough to effectively communicate what must be done. So now we must apply a few WBS concepts simultaneously to a real-life project. This is where Use-Related Characteristics come into play. Let's use our house metaphor to show what we mean (see Exhibit 4.2).

The house metaphor shows the breakdown of the WBS at levels 1 through 4. Taking a closer look, we see that only one of the "legs", or branches of the WBS is decomposed to level 4–and that is the Primary Structure. The other major elements of the WBS haven't been broken down at all, and remain at level 2.

A common way to determine how many levels there are in a WBS is to add up how many individual characters (separated by periods) there are in the numbering scheme associated with a specific WBS component. For example, component 1.1.1.3 in Exhibit 4.2 is at level 4 of the WBS. This can be determined by counting the individual characters (excluding the periods), as shown in Figure 4.2.

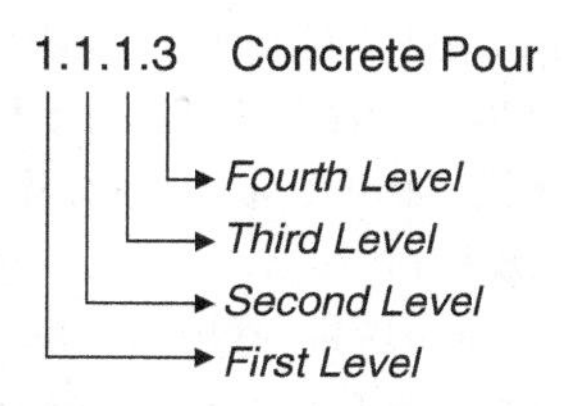

Figure 4.2 Level Number Illustration.

Is this controversial or confusing? We say "No, it isn't." Here, we have decomposed the Primary Structure because it appears early in the project. Certainly, if this were a real project, we wouldn't end the development of the WBS there. We would surely further elaborate the Electrical, Plumbing and Inside Wall development–because there would be quite a few other elements in the WBS, and those would probably not be broken down early in the project. For instance, "Landscaping" and "Walkways" would naturally appear in the fully elaborated House Project WBS, but those elements may not need to be fully decomposed at the start of the project. In fact, if any of you are like us, the later this kind of thing gets locked down the better.

Eric speaking here: My wife can't "envision" things in her mind's eye as easily as I can–so if this were my home being built, detailing the landscape and walkways at the beginning of the project would only serve as a starting point for the multiple changes she would ask that we make after she's seen the house standing on the lot. And if you've ever had anything like a home built for you, you know that every single one of those requests comes in the form of a formal change document to the contractor–with its attendant Cost Estimate... (ouch!). So to avoid the inevitable, we would delay elaborating the landscaping. Your situation may be different–and if so, you would decompose this at a different time in the project, earlier perhaps.

Robert speaking here: My wife is the opposite. As a professional architect, she does "envision" work in her mind's eye usually before I can.

This, in essence, is an example of a Use-Related Characteristic. The WBS must be designed in such a way as to ensure it meets all the needs of the project and project stakeholders. For Eric, this elaboration would be delayed to occur late in the project, for Robert's wife or you, possibly early in the project. "It depends" is the best way to describe whether this project element should be done early or late–it depends on the **use** of the WBS within the performing organization and project.

This is only one example of a Use-Related Characteristic. There are many other Use-Related Characteristics, but they are too numerous to discuss here. The overarching concept is that if we accept the notion that "the quality of a WBS depends on how well the specific content and type of WBS elements meet all the needs for which the WBS has been developed" (*Practice Standard for Work Breakdown Structures*–Second Edition, p. 20) is true, then Use-Related Characteristics that apply varying degrees of detail and decomposition depend on the needs of the project–and ensure the WBS meets those needs.

Here are areas where the use of the WBS drives its construction. Depending on the circumstance, the Project Manager would develop a very detailed WBS or a very simple one. Additionally, he or she may fully elaborate the entire WBS at the start of the project or save the elaboration of some components for later phases. Whatever the case, the Project Manager would be careful to develop a WBS that meets his or her needs for managing the effort. It is very easy to create deeply detailed Work Breakdown Structures for projects that simply overwhelm the Project Manager and team. We have seen Work Breakdown Structures developed for projects that have literally thousands of elements in them. In many of these cases, we find the WBS on a shelf somewhere or tucked in a file. Though these may be accurate and very detailed depictions of the work of the project, if the WBS isn't being used to manage the project, it is of no value. The objective is to rely on the WBS for project guidance and decision-making. If it can't be used for that purpose, there isn't much use for it at all.

Beyond this, the WBS should be developed in a way that ensures it achieves a sufficient level of decomposition for controlling the project, for communicating the work of the project effectively across and throughout the project organization, for providing stakeholder oversight and for assigning accountability at the work package level. How decomposed should this be? Again, "It depends." It depends on the nature of the project and of the project organization.

Consider this: If you worked in an organization where a particular process was used to deliver products time after time, the WBS for the product delivery would likely be similar from project to project. Some elements would be modified to meet the needs of certain projects, but in general, the various Work Breakdown Structures for the organization would probably look very similar if you were to compare one project to another. A pharmaceutical company's Work Breakdown Structures would likely fit this model, as the processes for researching, developing, receiving approval for and introducing new pharmaceutical products does not change significantly from product to product. These processes are generally static and repeatable. Moreover, in an organization where the project processes are repeated, familiar and known to the project teams, the Work Breakdown Structures can be less elaborate, because there is less of a need to detail what a majority of the project team members know.

By contrast, if each project an organization takes on is new and different, the WBS created for the projects would need to be specifically designed to meet the needs of individual deliverables and product outcomes. Hence, each of the Work Breakdown Structures in this organization would be unique and have varying levels of detail, dependent on the project's needs. As you can see, there is no single answer to the question of how much decomposition is appropriate–the answer is "It depends."

To recap this discussion, we would like to encourage you to begin with the WBS–and determine the appropriate level of decomposition (Use-Related Characteristic) ensuring all elements are represented in the WBS–but most importantly, modify the WBS as necessary to make it fit your unique circumstances. If, on the one hand, you or your stakeholders require a very detailed WBS and want to manage from a very detailed document, produce it. On the other hand, if your stakeholders aren't familiar with WBS construction and don't fully understand how to apply the WBS, then don't overwhelm them with elaborate detail. Provide a breakdown of the work that speaks to project team members and stakeholders at their level of understanding. Don't force the WBS to be highly elaborated when the situation doesn't call for it. Once you are comfortable developing and applying Work Breakdown Structures, allow the needs of the project and organization to dictate the level of complexity and detail for the WBS you produce.

WBS DICTIONARY

As the WBS is created for the project, the corresponding WBS Dictionary should also be developed to support it. While many practitioners do not feel this is a "must have" and often avoid creating the WBS Dictionary,

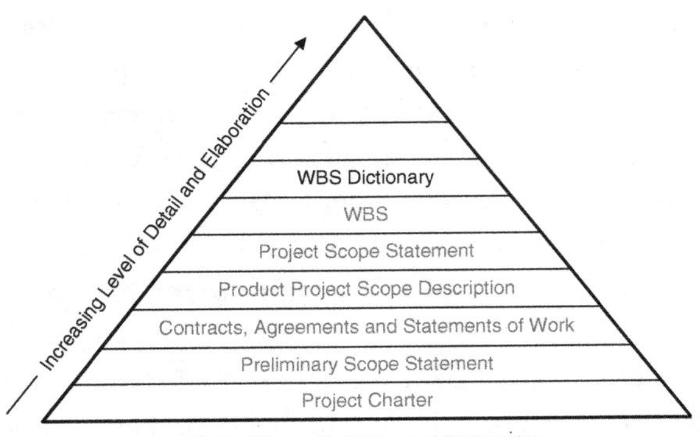

Project Scope Definition and Elaboration

we have discovered that the WBS Dictionary provides an additional level of clarity for all team members, sponsors and stakeholders. By providing the necessary explanations, context and detail the WBS Dictionary aids communication and facilitates understanding. Your authors believe this "supporting" project element is an evolving standard practice, and we strongly recommend and encourage all Project Managers who create Work Breakdown Structures for their projects to include the WBS Dictionary. This will provide critical information for everyone directly participating in the project or those touched by it–including sponsors, stakeholders and customers.

As a reference point, the word *dictionary* is defined as "A reference book containing an alphabetical list of words, with information given for each word, usually including meaning, pronunciation, and etymology." (American Heritage Dictionary, www.dictionary.com). Utilizing this definition in a very literal sense, you can expect the WBS Dictionary to be an explanation of each of the elements within the WBS. The WBS Dictionary is a further elaboration of what is included in each element within the WBS, including boundaries or scope for each. As a result, the WBS Dictionary provides a clear definition of the project's deliverables for all project team members, and articulates specifically what each element within the WBS is expected to address. The WBS Dictionary can also aid in the resolution of questions regarding scope that can not be explained with the simple WBS element entry, and will aid in communication to the entire team. As stated, the authors believe that the WBS Dictionary is nearly as important as the WBS itself and adds significantly to the effectiveness of the WBS for the management of a project.

Speaking of the management of the project, now is a good time to discuss a couple of different approaches to managing the project. We can now address the question about managing to an effort based on activities as opposed to managing the project based on deliverables. We had originally switched these items in the outline for the book–speaking about activity-based and then deliverable-based management. As part of the research for this book, we chose to discuss deliverable first and then activity-based–they seem to flow better this way, and as you can see, we want to emphasize deliverable orientation. These two are compared and contrasted in the next sections.

DELIVERABLE-BASED MANAGEMENT

With **Deliverable-Based Management**, the control concept is taken beyond the Work Package to tasks and activities in the Project Schedule. Here, Work Packages are decomposed into individual tasks and activities, the outcomes of which are finer-grained deliverables. When a Project Schedule for this approach is created, each task and activity will produce a "deliverable" that can be combined with deliverables from other tasks and activities to form the intended Work Package deliverables. And as we've discussed previously, Work Packages naturally roll up into higher-level WBS elements. With this approach, the entire project is managed by the creation and integration of these individual and compound deliverables. The ability to manage the project at the detail level and roll up the deliverables based on the WBS hierarchy is the true power of deliverables-based management. We strongly believe that a deliverable-based approach is what should be used for most projects, most of the time.

Deliverable-based management allows for schedule, cost, resource and quality to be understood, aggregated, measured and monitored at both a specific deliverable and higher-level WBS element level, thus providing the Project Manager with the ability to visualize and communicate (particularly for dependent deliverables) how the project is performing. In addition, if one deliverable is not being achieved, this methodology allows the Project Manager to quickly perceive the impact this deliverable has on others, as well as on the project as a whole. This level of awareness and control is difficult to achieve when the project is managed using Activity-Based methodologies. This approach, if followed, provides a much-needed level of traceability throughout the entire project.

ACTIVITY-BASED MANAGEMENT

In contrast, **Activity-Based Management** (ABM) is a valid technique for use in support of ongoing business operations. In these instances, cost at the activity levels is essential to managing and controlling the costs of an overall business. With ABM, the manager can tie costs (direct, indirect, and overhead) to specific operational activities, and is able to control associated costs as necessary. While this is appropriate for the ongoing operations of a business, it is often inappropriate for the management of a project or program. It is often difficult, time consuming, and costly to perform a detailed analysis related to specific operational deliverables because these activities typically impact several deliverables simultaneously and may cross business units or functions. It is precisely this inability to group work and cost against the WBS hierarchy to show traceability that makes this type of management problematic for the Project Manager.

As you can see, both Deliverable and Activity-Based Management have their appropriate uses. While it is not always easy to make this determination, it remains important to utilize the appropriate methodology for the appropriate situation.

SCOPE BASELINE

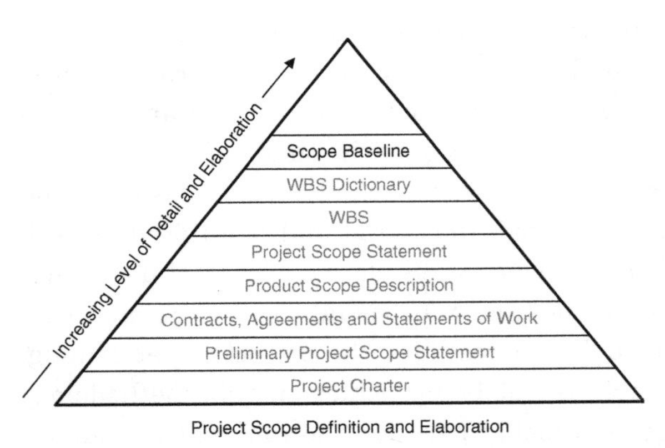

After the WBS and the corresponding WBS Dictionary have been completed, the triumvirate of **Scope Baseline** is complete. This includes the project Scope Statement, the WBS and WBS Dictionary, and these items become the baseline against which all changes to the project will be managed and measured. Once these items have been approved, agreed-upon and baselined, formal change management for the project begins

ACCEPTANCE CRITERIA

One last item to cover in this chapter is **acceptance criteria**—that is, how will the acceptance or declination of a project and its results be determined? Utilizing the WBS and its defined deliverables, the project team can not only drive to the preferred outcomes, but also determine the appropriate acceptance criteria against which the project should be measured. This will allow the project team to define early on how the *product* scope (remember, features and functions) will be accepted, and also how the deliverables in the WBS will be accepted.

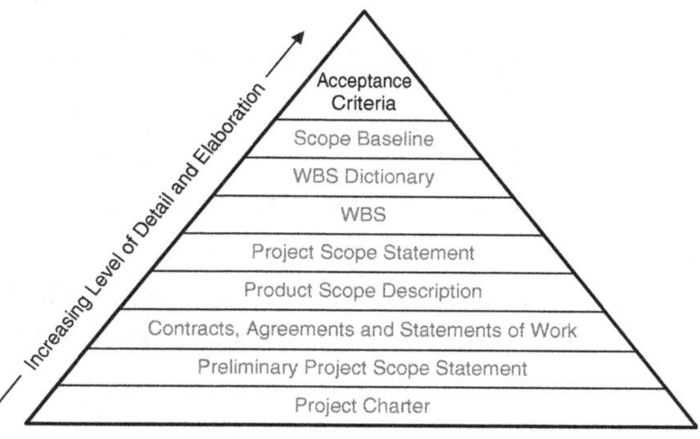

Project Scope Definition and Elaboration

If the acceptance criteria are *not* defined in the WBS Dictionary, there are a couple of issues that could be encountered. First and foremost, in absence of acceptance criteria, project team members responsible for creating project deliverables have no target to drive toward. The acceptability of the individual output is then left to chance. Will the deliverable the team member has produced be acceptable to the customer? Perhaps yes, perhaps no. Without explicit acceptance criteria, this becomes a negotiation between provider and customer (buyer and seller) rather than a yes or no acknowledgment. This can lead to questions about the degree of completeness and quality across the entire set of project deliverables and potentially can result in an unsatisfactory customer experience. Beyond this, without a defined set of acceptance criteria tied to the WBS and WBS Dictionary, the Project Manager will have a difficult time determining when the project is truly complete and ready for turnover to the customer.

As stated previously, without specific acceptance criteria, customer acceptance and transition of the project can be a moving target. When this occurs, it can delay turnover to the customer and impacts the start of the warranty period (should this apply). Finally, the absence of acceptance criteria can impact the closeout activities for the project and may ultimately prevent the project from being fully and formally closed. Exhibit 4.3 shows (an example of) the acceptance criteria for our House Project.

While these are not the only planning activities within a project, we have reached the point at which the project is baselined and the WBS and associated artifacts are placed under the control of Change Management. In the next three chapters, we will discuss additional planning activities that occur after the WBS has been created and baselined. We will also examine how these processes interact with the WBS.

House Project–Buyer's Acceptance Criteria

- All work completed and approved–per contractor's signature(s)
- All internal punch list items resolved
- All external punch list items resolved
- All systems tested and perform according to specifications
- All inspections completed and passed
- House is cleaned and "move-in" ready

Exhibit 4.3 House Project Acceptance Criteria

CHAPTER SUMMARY

In this chapter we have answered the questions of why, who, what, when, where and how regarding the interaction between the Work Breakdown Structure and the applicable project scope planning processes.

We began the chapter discussing the next two artifacts to be completed in any project–the Product Scope Description and the Project Scope Statement. The Product Scope Description allows the team to define and elaborate on the specified features and functions that will be included in the final product(s) delivered to the customer. At the same time the Preliminary Scope Statement is revised to reflect the updated view of what the project will deliver as well as the work needed in order to produce the desired deliverables. This becomes the Project Scope Statement.

Once these artifacts have been completed, our attention turns to the creation and delivery of the Work Breakdown Structure. In developing the WBS, we have answered the following questions:

Why: It is critical to complete the WBS in order to define and detail all of the work that must be completed in order to complete the project objectives.

Who: The WBS should be completed by the project manager and team. It is their responsibility to not only build, but also maintain the WBS during the project's lifecycle.

What: Included in the WBS should be the scope of the project and only the scope of the project. Nothing more, nothing less. A good rule of thumb is if it isn't in the WBS, it should be not be created or delivered as part of the project.

When / Where: Early and continuously. As we have stated, the WBS is a critical element to be utilized in defining and maintaining the scope of any project. In addition to this, it should be utilized to measure the progress of the project toward completion of the deliverables and objectives.

How: This is the largest portion of this chapter and includes the steps to create the WBS. These include starting with what is known from the Project Charter, Preliminary Scope Statement, contracts, agreements and Statements of Work. From there, the tools available and the method of creating the WBS are determined. These can vary from pencil and paper to a software tool and from the top-down to bottom-up methods. In parallel, the WBS Dictionary should be created to enhance and further define the elements within the WBS. Once the WBS has been built, the Core and Use-Related Characteristics defined in Chapter 2 are utilized to ensure the level of quality within the WBS. Finally, the WBS is placed under the control of Change Management to control and manage changes to the project's scope throughout its lifecycle.

The next section of the chapter discusses the Deliverable and Activity-Based methods of approaching and managing the project–outlining the advantages and disadvantages of each method and where it is best utilized.

The final section of the chapter discusses Acceptance Criteria. These can easily be built directly from the elements in the WBS and its corresponding WBS dictionary. Examples for the House Metaphor were provided in the chapter.

As you can see from this information, the WBS is essential to the success of a project—not only during the scope elaboration and planning phases, but also during the execution, management and control of the project during its entire lifecycle.

REFERENCES

Definition of "Dictionary". Available from the American Heritage Dictionary at www.dictionary.com.

Project Management Institute. 2001. *Practice Standard for Work Breakdown Structures*. Newtown Square, PA: Project Management Institute.

Project Management Institute. 2004. *A Guide to the Project Management Body of Knowledge (PMBOK® Guide*—Third Edition). Newtown Square, PA: Project Management Institute.

CHAPTER QUESTIONS

1. Which of the following project documents describes the look and feel of the outputs of the project in narrative format?
   a. Project Charter
   b. Preliminary Project Scope Statement
   c. Product Scope Description
   d. Work Breakdown Structure

2. Quality Work Breakdown Structures include which of the following? (Select all that apply.)
   a. Internal deliverables
   b. External deliverables
   c. Interim deliverables
   d. Ad hoc deliverables

3. Which of these WBS creation approaches involves first defining all of the detailed deliverables for the project?
   a. Top-down
   b. Bottom-up
   c. WBS standards
   d. Templates

4. Quality Work Breakdown Structures must have at least ________________ levels of decomposition.
   a. One
   b. Two
   c. Three
   d. Four

5. Which approach to the management of projects allows for schedules, cost, resource, and quality to be understood, aggregated, measured, and monitored at both a specific deliverable and higher-level WBS element level?
   a. Activity-based management
   b. Task-based management
   c. Deliverable-based management
   d. Milestone-based management

Chapter 5

The WBS in Procurement and Financial Planning

CHAPTER OVERVIEW

This chapter delves into two key uses of the WBS during the Planning phase. The opening section of this chapter discusses how the WBS supports the build versus buy decision-making process. The concluding sections of this chapter examine the role of the WBS during the Cost Estimating and Cost Budgeting processes.

The major sections of this chapter include:

- Build versus Buy Decisions
- Cost Estimating
- Cost Budgeting
- Cost Breakdown Structure

BUILD VERSUS BUY DECISIONS

As we have discussed previously, the WBS contains the complete deliverable hierarchy for the project or program. One of the many benefits of deliverable-oriented Work Breakdown Structures is that it facilitates the **build versus buy** decision-making process. Within a deliverable-oriented WBS, individual work packages (deliverables) or entire portions of a WBS hierarchy can be evaluated to determine whether those deliverables should be built by the project team or purchased (contracted or sub-contracted). If a buy decision is made,

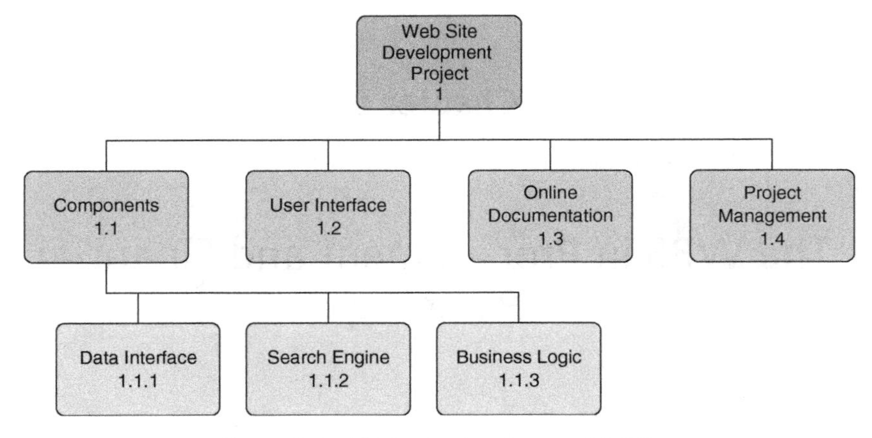

Figure 5.1 Web site development WBS.

further evaluation may be required to determine whether a purchase, lease, or rental is more appropriate to the needs to the project.

Figure 5.1 depicts a portion of a WBS for a Web Site Development Project. To deliver what is required as defined in this WBS, the project team will need to implement all of the components that fall under Components (1.1). One of the components of the Web site will be a search engine. The project team can either internally develop a search engine or choose to contract externally for that component.

After some analysis, the Project Manager decides to procure a search engine available on the open market. Here again, many options exist. Among them include (1) a search engine company can be purchased outright (not necessarily a good option, but an option nonetheless), (2) a search engine module can be licensed and embedded in the Web site, or (3) a search engine can be obtained from the open-source market and embedded in the Web site. For the purposes of this example, we will say that option 2 is selected. In this simple example, WBS element 1.1.2 will be procured from a third party external to the delivering project organization. From a project delivery standpoint, the deliverable has not changed—1.1.2 is still the search engine component of the Web site. The only difference is that the WBS element will be assigned to a third party provider, not an internal project team member.

With a high-quality WBS and WBS Dictionary, enough information about each of the deliverables is available to enable the project team to evaluate the relative advantages and disadvantages of building the deliverable versus buying it. With the proper information available, the project management team can determine which deliverables are within the capabilities of the direct delivery organization(s) and which must be procured from outside resources. If a buy decision is reached, the project team can setup a new quality process or utilize an existing one to help ensure that the best possible outcomes for the project are delivered by the most qualified resources.

The *PMBOK® Guide*—Third Edition recognizes the importance of the WBS and the WBS Dictionary in build versus buy decisions as they are included as key inputs to Plan Purchases and Acquisitions process (*PMBOK® Guide*—Third Edition, p. 54). In fact, build versus buy decisions are a key input or output of several of the Procurement related processes in the *PMBOK® Guide*—Third Edition.

The WBS can also be used as a vehicle for facilitating sub-contracting work and establishing individual deliverables owned by specific providers. Values can be place on the individual deliverables and included in the WBS Dictionary. This information can then be utilized and monitored during the Execution and Monitoring and Controlling processes to help ensure quality and completeness for the sub-contracted deliverables.

COST ESTIMATING

Cost estimating involves developing cost approximations for each WBS work package, including any and all costs required to complete the work package deliverable(s). The cost estimates should include all *product* and project-related costs, such as the following:

- Resource costs
- Material costs
- Quality costs
- Risk response costs
- Communications costs

The information included in the WBS Dictionary can provide valuable insight into the Cost Estimating process. The WBS Dictionary can include scope boundaries for the deliverables, a cross-reference with a **Responsibility Assignment Matrix** (RAM) or even a detailed description of the deliverable that may include elements required in the costing process. This information can in turn provide valuable guidance for estimating of

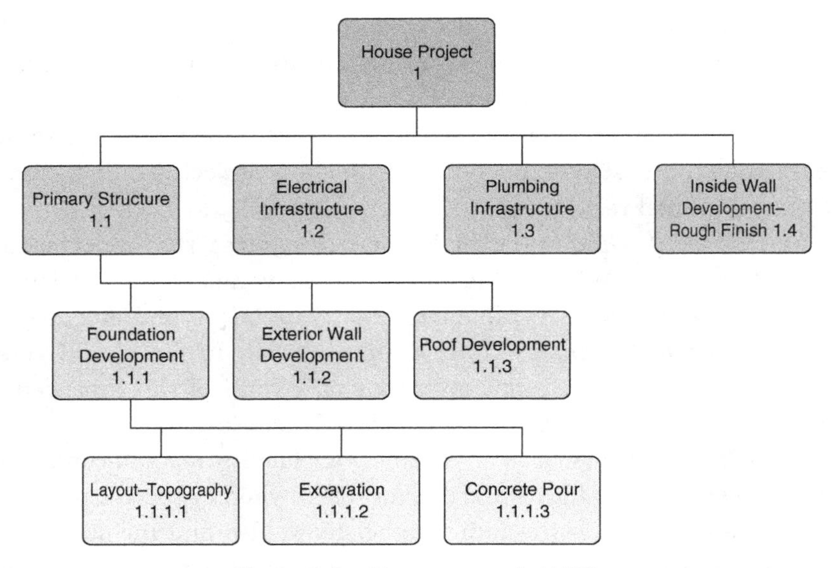

Figure 5.2 House example WBS.

the work package costs. The more defined the WBS Dictionary, the more refined the estimates.

Once again, let's use the House metaphor to illustrate this point. Figure 5.2 shows the Work Breakdown Structure for the House project. Our objective now is to estimate the costs for the three Work Packages that are part of the foundation development WBS element.

For this project, it was decided that the Layout—Topography element would be contracted externally from a third party firm that specializes in this type of work. Given this, the costs to the project are estimated to be $8,000 to the third-party provider plus another $2,000 in material costs. So the total cost of the Layout—Topography work package is estimated to be $10,000.

For Excavation, that work will be handled internally by the project team. The costs for this WBS element not only include labor and materials, but also the costs for required permits as well as rental costs for the heavy equipment that will be needed. When all is said and done, the estimate for the Excavation work package is $30,000 and is broken down as follows:

- Materials – $3,500
- Labor – $15,000
- Permits – $1,500
- Rentals – $10,000

The *PMBOK® Guide*—Third Edition recognizes the importance of the WBS and WBS Dictionary to cost estimating as they are considered key inputs to the Cost Estimating process (*PMBOK® Guide*—Third Edition, p. 51). Ultimately the deliverables defined in the WBS drive the estimating process, which in turn drives the budgeting process.

COST BUDGETING

Following the estimating processes for each of the work packages, **Cost Budgeting** is used to consolidate and aggregate individual work package estimates into an overall cost for the project—and to establish the initial cost baseline. The aggregation of costs at the work package level can and should follow the hierarchical construct of the WBS itself. Structuring the costs in this manner provides several advantages including:

- The 100% Rule used in the creation of the WBS hierarchy ensures that each level of decomposition includes 100% of the parent elements. Following this construct for cost estimating ensures that the cost budget will ultimately include 100% of the costs, assuming that the WBS is completely and correctly defined.
- The WBS elements in the hierarchical structure are used as the Control Account, thereby ensuring synchronization between how the work is defined and how it is performed and managed.
- The WBS provides for roll-up of costs, similar to deliverables roll-up in a WBS hierarchy.

Figure 5.3 shows our House example with estimated costs included within the individual WBS elements. Now for those of you who are saying that these costs are unrealistic, we want to state here and now that we have not attempted to make the cost illustration realistic to scale. We are using this simply as an example, and we request that you kindly suspend the disbelief for a moment. And for those of you wondering whether this example refers to U.S., Canadian or Australian dollars, the answer is yes. Please remember that the numbers included in the example are for illustrative purposes only.

As you can see from Figure 5.3, the estimated cost of the House is $180,000. This cost can be broken down into $100,000 for the Primary Structure (1.1), $15,000 for the Electrical Infrastructure (1.2), $25,000 for the Plumbing Infrastructure (1.3) and $40,000 for the Inside Wall

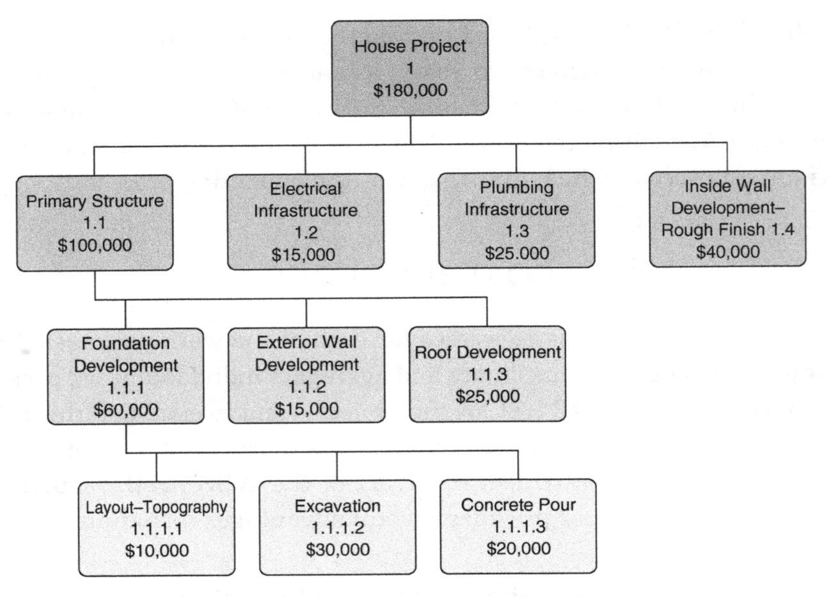

Figure 5.3 Cost budget—WBS integration.

Development—Rough Finish WBS element. The Primary Structure in turn is broken down further, as is the Foundation Development. As you can see, the costs from WBS elements 1.1.1.1, 1.1.1.2 and 1.1.1.3 roll-up equal to the cost of WBS element 1.1.1. This roll-up can be seen all the way up to the root of the tree, $180,000 for the House project.

The *PMBOK® Guide*—Third Edition defines a control account (formerly called a cost account) as "a management control point where the integration of scope, budget, actual cost, and schedule takes place, and where the measurement of performance will occur" (*PMBOK® Guide*—Third Edition, p. 355). This concept of using the WBS as the baseline for project control is embedded throughout the *PMBOK® Guide*—Third Edition. This is why we say that the WBS is fundamental to the management of cost and time, not just scope.

COST BREAKDOWN STRUCTURE

A **Cost Breakdown Structure** (CBS) is a hierarchical breakdown of cost components for a project. Like a Work Breakdown Structure, it can be

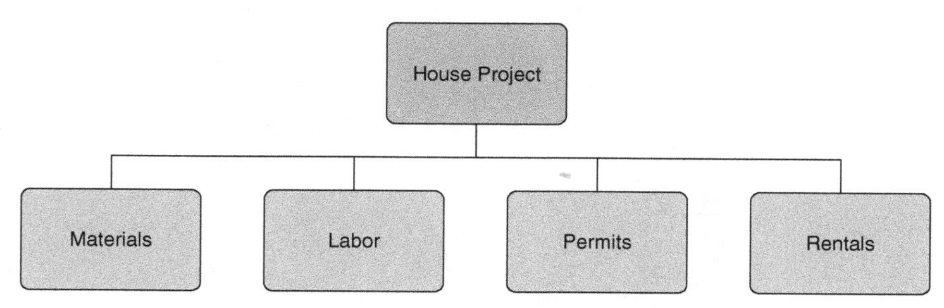

Figure 5.4 Cost breakdown structure.

depicted hierarchically. Figure 5.4 reflects a Cost Breakdown Structure for the House project. Here again we ask that you suspend disbelief, as this is only an illustration. In this simplified example, the CBS reflects four types of project costs—materials, labor, permits, and rentals.

Having a completed WBS, Project Budget and CBS, it is now possible to integrate these three tools and depict the cost breakdown of individual Work Packages, as shown in Figure 5.5 (see page 82).

This example shows the WBS and Project Budget further integrated with the CBS. To keep the illustration simple, we have only depicted the three work packages for the Foundation Development WBS element of the House example. Integrating the use of these project management tools provides powerful insight not only into the breakdown of the project's scope but also its costs. It also provides a unique viewpoint for presenting what could be complex information to the project stakeholders in easy to understand manner. This is all made possible by aligning these key project planning tools and processes with the project's Work Breakdown Structure.

CHAPTER SUMMARY

This chapter addresses some of the planning aspects of Work Breakdown Structures. The initial section discusses how the WBS aids in build versus buy decision making. The next two sections explain how the WBS is an integral part of project cost estimating and budgeting. The chapter concludes with a discussion of Cost Breakdown Structures and provides an example of the integration between a CBS and WBS.

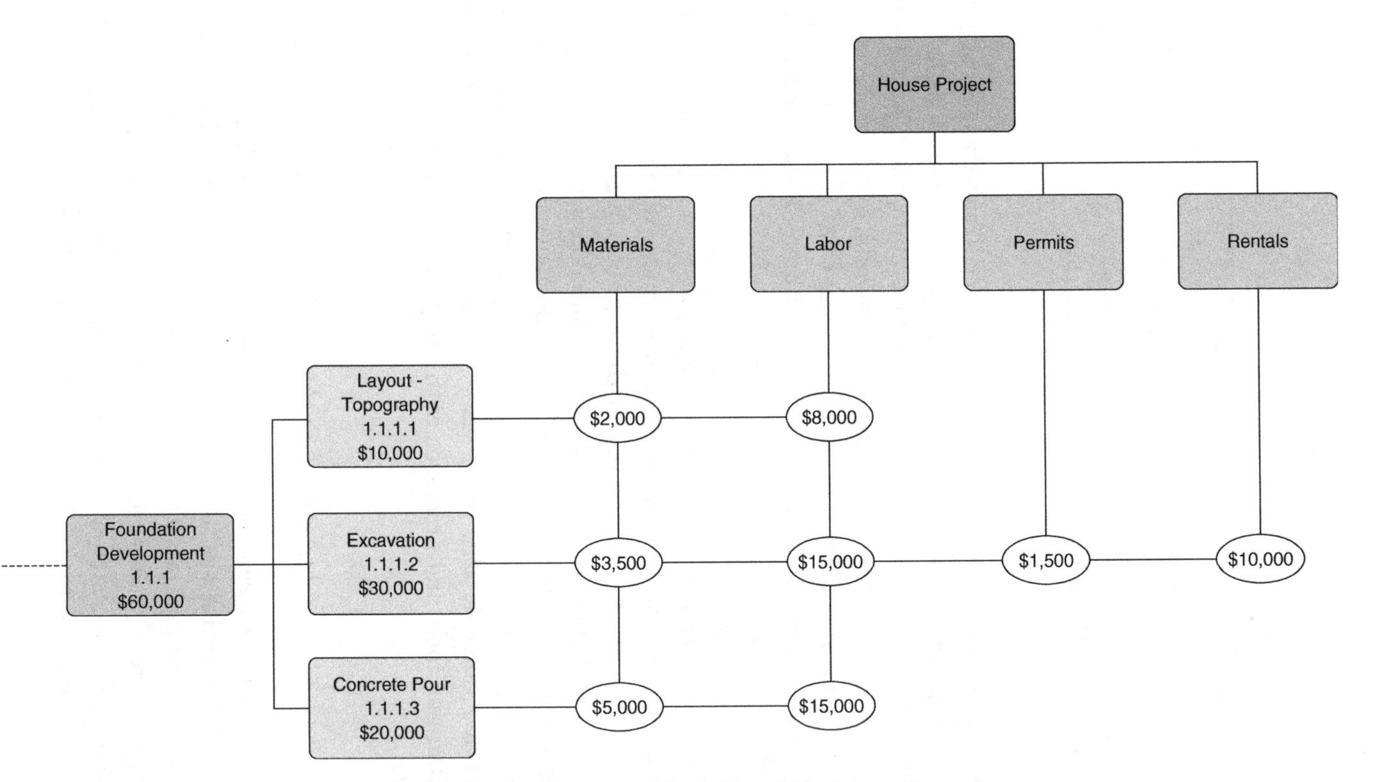

Figure 5.5 WBS, project budget and cbs integration.

REFERENCE

Project Management Institute. 2004. *A Guide to the Project Management Body of Knowledge (PMBOK® Guide)*—Third Edition. Newtown Square, PA: Project Management Institute.

CHAPTER QUESTIONS

1. Work Package cost estimates should include which of the following? (Select all that apply.)
   a. Resource costs
   b. Material costs
   c. Quality costs
   d. Risk Response costs
   e. Communications costs
   f. a and b only
2. Which scope-management-related deliverable can greatly aid in cost estimating?
   a. Project Charter
   b. Project Scope Statement
   c. Product Scope Description
   d. WBS Dictionary
3. Which of the following are advantages for structuring cost budgets following the construct of the WBS? (Select all that apply.)
   a. The 100% Rule used in the creation of the WBS hierarchy also ensures that the cost budget will ultimately include 100% of the costs.
   b. The WBS elements in the hierarchical structure are used as the Control Account, thereby ensuring synchronization between how the work is defined and how it is performed and managed.
   c. The WBS provides for roll-up of costs, similar to deliverables roll-up in a WBS hierarchy.
   d. b and c only
4. What is a "management control point where the integration of scope, budget, actual cost, and schedule take place"?
   a. WBS Element
   b. Work Package
   c. Control Account
   d. None of the above
5. What is a hierarchical breakdown of project cost components?
   a. Work Breakdown Structure
   b. Resource Breakdown Structure
   c. Organization Breakdown Structure
   d. Cost Breakdown Structure

Chapter 6

Quality, Risk, Resource and Communication Planning with the WBS

CHAPTER OVERVIEW

As the planning for a project evolves, additional project management processes come into play. These familiar processes are vital to the project's forward progress and help ensure the delivery of the project's agreed-upon outcomes—the products, services, or results derived from the effort. In this chapter we discuss the interaction between the WBS developed for the project and these core project processes including the planning activities for Quality, Resource, Risk and most importantly, Communications.

This chapter addresses the following topics:

- Approaching Quality, Resource and Risk Planning
- Using Existing Templates and Processes
- Creating Processes to Support the Project
- Using the WBS as a Basis for Process Development
- Employing the WBS and WBS Dictionary
- The Whole is not Greater than the Sum of Its Parts—It Equals Precisely 100% of the Sum of Its Parts
- Examining Process Considerations
- Communications Planning Using the WBS as a Foundation
- Developing the Communications Plan including the Communications Matrix, Hierarchy of Information and the Meeting Matrix

For each of these processes, the WBS plays a vital foundational role:

- Because the WBS details each of the deliverables of the project, we rely on it during **Quality Planning** to help determine which quality processes are appropriate for individual project deliverables. Beyond this, the WBS and WBS Dictionary provide clarity regarding the project's scope, enabling the project manager to develop Quality Assurance Plans and activities that detail quality processes for the project. Finally, as the project execution phase gets under way, Quality Control processes are implemented to ensure the project's deliverables conform to quality criteria for acceptance. During this phase of the project, continual monitoring and control efforts are tied to quality checkpoints that enable the project team to sample various interim outputs and make adjustments where necessary. We will examine how the WBS performs its role in the identification and measurement of appropriate controls.
- **Resource Planning** involves mapping specific resource needs, identified through decomposition of individual work packages to the project's scheduled tasks and activities and allows integration of the project's RBS (Resource Breakdown Structure), OBS (Organizational Breakdown Structure) and WBS. Here, we demonstrate how the WBS supports the alignment of resources to specific needs within the project.
- **Risk Planning** is one of the few forward-looking project management processes that can provide insight about what the project team might encounter during the execution of the project. Like project estimates that are based on the anticipation of *need*, risk planning looks at potential future *events* that may impact the project and helps the Project Manager and team develop effective plans for dealing with these events, should they occur. The WBS provides a vehicle for perceiving and describing risks that may positively or negatively impact the project's various work elements and deliverables. In fact, given that effective forward-looking risk planning deals specifically with individual project deliverables, the knowledge that the work of the project is itself decomposed into individual deliverables (as a product of the WBS) compels the project manager to ground Risk Planning on the WBS developed for the project. The WBS and WBS Dictionary together provide a clear view of the project's scope and boundaries—and provide a ready tool for the Project Manager to

enable the identification of risks and development of effective risk response strategies.

- Finally, the success of **Communications Planning**, one of the primary roles of the project manager, relies on the project manager's thorough and clear understanding of the communications needs of each of the project's stakeholders and stakeholder groups. In this chapter we share strategies for the development of overarching communications plans that simplify this process while tailoring communications methods and modes to the specific needs of individual stakeholder groups.

APPROACHING QUALITY, RESOURCE AND RISK PLANNING

During the critical, initial phases of a project, Initiating and Planning, the project manager is faced with a crushing set of responsibilities and challenges. Confucius said, "A man who does not plan long ahead will find trouble at his door." He or she must demonstrate strong leadership for the current members of the project team as well as the various stakeholders and stakeholder groups, and in most cases, set tone and direction when a great deal of information about the project remains unknown.

Often at this stage requirements are vague, with short statements composed of simple sentences containing little detail. These "requirements" statements may capture only high-level functions, use cases, attributes or outcomes such as "the product must conform to applicable legal and regulatory constraints" or "the building may not exceed 91 feet above street level." At the same time, designs are often incomplete and provide little insight into the complexities of the work ahead, as they have typically been developed at the highest level to facilitate communicating (and "pitching") the overall product concept. As a result, they are a far cry from the detail required to support building the actual product or service.

The approach for build versus buy decisions that might apply to certain project components has not yet been determined and it is quite likely that many of the resources required for the heavy lifting of the project have not yet been identified. With so much yet unknown, how does the project manager move forward to effectively plan key processes for the project? Fortunately, there are two ways to approach answering that question. The choices are simple. The project manager can either apply the existing processes within the company, division or organization to meet the project's needs for managing Quality, Resourcing, Risk and Communications, or he/she can create them where they are needed. That's all there is to it—either use what is there or create the necessary processes to support the project. Moving forward without these processes in place isn't an option—the project will be fundamentally flawed without them, so the choice is truly quite simple.

We'll wager, however, that right now you're thinking most of the time these core project management processes simply don't exist, so you *must* create them whether you wish to or not. And you would be right, but only a percentage of the time depending on the industry or sector we're discussing. In some instances—for example in aerospace, the pharmaceutical industry, the United States Department of Defense and travel—these processes do exist and have been in use supporting key project initiatives for quite a long time. In fact there is evidence (for example, U.S. government project document filings) in these sectors that key processes have been in place long enough to become very sophisticated, detailed and refined. In other sectors, however, such as in new technologies and parts of the information technology industry for example, these processes are rarely stable enough to call them "repeatable." Project Managers in these sectors find themselves creating the same core project management processes over and over again.

Let's now examine these two paths a little more closely. Later in the chapter, when we look at creating repeatable, sustaining-type processes for projects rather than using the existing processes within an organization or company, we will discuss how the WBS performs its role during process development and implementation. As an additional note, we'll address Communications Planning separately, because the uniqueness and critical role of communications in all projects warrants addressing the development of effective communications processes as a separate subject. As we have mentioned previously, at this particular point in the project's lifecycle the project manager finds himself or herself at a crossroads and can follow two very different approaches to setting the necessary foundational processes in place. Regardless of the direction taken, however, it's clear that now is not the time for the project manager to determine how to move forward (though this is often when the decision is made). This planning is something that might have been better determined before the pressure was on, before all eyes were on the project and project manager. Preparing for the implementation of core processes to support a project takes a degree of research and planning on the project manager's part, and the time necessary for this work must be anticipated or "planned in" as part of the effort.

USING EXISTING TEMPLATES AND PROCESSES

One sure path to success for the project manager is to apply tried and tested processes for Quality, Resourcing, Risk and Communications. At the outset this assumes that the processes currently exist and are available to the project manager for use. These are processes that would have evolved over time within the division, company or organization as a result of repeated use—and would have proven themselves to be efficient and effective. Of course, this requires and implies that the institution or organization where this occurs has invested in establishing and maintaining these processes, and has done so purposefully, and most likely, out of necessity. The products, services and results these companies and organizations provide must meet rigorous quality standards and must do so with a degree of reliability and consistency that is difficult to attain.

Imagine for a moment the airline industry. At the time we are writing this book and over the recent past, we have witnessed a near meltdown in the entire airline industry's ability to keep to scheduled flight departure and arrival times. To takeoff or land at your flight's advertised, scheduled times, or to experience flights where both occur as scheduled is becoming a rarity. Public dissatisfaction and outcry is on the increase, the airlines and airports are under fire. and the problem has attracted the attention of both U.S. and international government agencies. Interestingly enough, however, the general traveling public hasn't stopped flying because of this problem. Sure, travelers are clearly annoyed and are speaking out in record numbers. The airlines are doing "what they can," but the truth of the matter is that this significant deviation from the implied standard for flight departure and arrival, or the poor schedule time "quality," is not a controlling factor for influencing public action. Because the benefits of flying, even with the recent inconveniences and problems, still outweigh the negatives and the time involved with finding alternative travel options.

Flight safety, by contrast, is an entirely different matter. The "quality" rules that apply to public safety in the airline industry are stiff and getting stiffer. (And the importance of this aspect of airline travel is likely causing some of the delays we've just discussed.) People would simply stop flying altogether if the "quality" attribute for safe flights were to be relaxed by no more than a few hundredths of a single percentage point. Here is what we mean: If we were to relax our existing quality control processes for airline safety relating to aircraft maintenance, flight crew off-time hours, flight crew alcohol consumption, passenger carryon baggage inspection, baggage or passenger screening, and we were to allow airline safety "incidents" (accidents) to suddenly jump from their current level to a number that is statistically represented by one standard deviation; meaning the difference between 6 sigma and 4 sigma, or an increase in likelihood of occurrence by 235 percent, people would likely opt for other modes of travel rather than to accept the risk of being on one of the less rigorously monitored and controlled flights.

In this case, project managers in airline industry operations typically apply existing processes for quality, risk, resource management and communications. Why? Because they must do so to ensure consistent and repeatable levels of quality, reducing known risk and verifying safe and secure travel for the public.

Also in this case, the airline industry as a whole has adopted a rigorous safety inspection and certification protocol. It consistently validates and verifies that all aspects of flight management meet specific safety standards. The individual airline companies have addressed this need and have developed standing processes to ensure the desired outcome. Reliable, consistent delivery of outcomes (results) by its very nature demands stable, refined processes to ensure high quality and fitness for use.

To draw another analogy, think now of production-line operations in companies or institutions that generate consumer goods such as appliances, vehicles, food products, tools, machinery and the like. In these cases, though the projects within the companies or organizations may be new, the project manager utilizes the existing available processes and resources such as the product quality assurance process, the administrative "engine" or staffing platform for filling necessary roles on specific project teams. The project may employ existing statistical modeling, sampling and analysis for quality assurance and use particular control techniques for measuring and monitoring the quality of the delivered products. Whatever the case, creating these processes is not a consideration and would be completely unnecessary. These functions are seen as valuable, reusable corporate assets supporting an array of project efforts and are often sophisticated and efficient mechanisms for the company.

Let's briefly summarize the use of existing processes and templates that are at the project manager's fingertips in some organizations:

- Organizations that produce known sets of products or services that must meet certain predetermined quality standards or produce products and services that evolve through incremental change typically maintain a set of processes and templates the project manager may rely on to deliver particular projects within those product/service sets.
- For quality management there are often specific quality criteria, quality assurance processes, and quality control tools available that have been refined over time and utilized repeatedly to deliver the company's products and services. The project manager's role here would be to utilize them fully and to modify them only slightly or perhaps not at all to ensure the products/services conform to specified criteria.
- Utilizing standard company processes for staffing individual project tasks is most beneficial to the project manager in organizations where these processes are known and stable. The company or organization that delivers familiar products and services often develops standing methodologies for addressing project resource needs and has "planned in" the use of employees, contractors and consultants at various predetermined volumes.
- By the same token, Risk Analysis and Risk Planning in companies and organizations such as this may also rely on tested methodologies. It is very likely the project manager will have access to complete Risk Management templates and tools used to carefully guide risk identification, risk analysis and risk response planning. These templates and tools often take the form of fully elaborated guides that pre-identify the risk categories most meaningful (high impact/high likelihood of occurrence) to the company and include risks that have been seen repeatedly in previous project efforts.
- Depending on the industry, company and business sector, specific categories of risk that the company must assuredly address appear in the templates and tools. These may include risk categories such as environmental, legal, regulatory, or statutory risks; operational risks such as contracting or administrative; and catastrophic or unexpected, unpredictable risks, including weather, theft, power

interruptions, physical environment events (explosion, building collapse, flood, etc.), and acts of war. Whatever the case, when these templates exist, they provide strong guidance for the project manager and project team and provide insightful evidence of the road ahead. The project manager is well advised to make full use of these valuable tools—applying them systematically during the planning phase, modifying and tailoring them to meet the needs of specific projects.

CREATING PROCESSES TO SUPPORT THE PROJECT

There are many instances where Quality, Risk, Resource and Communications management processes such as those described previously simply do not exist. There are endless reasons for this and we could begin enumerating them here, but in all likelihood we would inadvertently overlook the one that describes your particular situation and you might not be interested in reading further if we were to do that, so we're not going to list them. Suffice to say that many of you, just as we have experienced ourselves, have arrived at this very same juncture, responsible for projects that must move forward with little more than sketched requirements and "back-of-the-napkin" engineering drawings.

Of course there are no existing processes that you can rely on to deliver the project—in some instances they haven't been necessary before this. And if these processes perhaps do exist, it's unlikely you believe they're robust or stable enough to support your project. Like it or not, you must create the processes you'll need for managing Quality, Risk, Resourcing and Communications, and you've got to do that immediately with little guidance.

You are aware you must also develop similarly effective processes for Change (Scope) Management, Schedule (Time) Management and Financial (Budget) Management, but those topics are not part of the discussion in this chapter, so we're not including them here.

UTILIZING THE WBS AS A BASIS FOR PROCESS DEVELOPMENT

If the Quality, Risk, Resource Management and Communications processes necessary to support the project do not exist, or are insufficient to effectively manage the project, the WBS provides a useful starting point for improving the situation.

We are aware that some project managers prefer to jump right in, to start process development by gathering the project team in an effort to begin brainstorming solutions to one or more of these challenges. If the objective is Risk Management, the team would begin brainstorming potential risks. If, on the other hand, the immediate need is Quality Planning or Resource Planning (staffing), the team would begin discussing potential quality checkpoints or gaps in technical or management skills. This is sometimes a profitable way to begin, because it can produce some immediate insight and may surface real challenges. This free-form, unstructured approach has a downside, however. In many instances the outcomes of this approach are hit-and-miss. The solutions found can address some obvious aspects of the processes effectively, while others may be completely overlooked or unintentionally left unaddressed. When this occurs, occasionally these oversights don't actually appear to be real problems until it is too late in the project to address them effectively. The project manager and team is then forced into a reactive mode, while the cost of correcting the oversight and filling the process need causes a significant impact to the project's forward progress.

In discussions with project managers about this, we've heard some rather interesting views about approach. Some claim there just isn't a better way to begin building solutions to this urgent issue than to bring the team together in an informal and unstructured setting to collectively or collaboratively brainstorm—the fewer the constraints on the brainstorming activity, the better. Others would opt for starting with whatever risk, quality, resource and communications processes the team members might have used in the past. The process development activity would include cobbling pieces of each together to form one cohesive process, or working with them to "hammer them into shape" to support the project.

Sure, these approaches may produce satisfactory results, but it is also quite possible that they may not. Considering project leadership roles, however, we must agree that the Project Manager needs a more reliable process development methodology for establishing key project processes early in the project's life. Moreover, rather than leaving this to chance, if we were to borrow essential guidance from Dr. Stephen Covey's *7 Habits of Highly Effective People*, perhaps a more effective approach to process development would be to "Begin with the end in mind." The WBS, with its Core and Use-Related Characteristics, provides a clear starting point for process development.

EMPLOYING THE WBS AND WBS DICTIONARY

As we discussed earlier, while it may be beneficial to brainstorm ideas about Quality, Risk and Resource Management, this unstructured approach introduces great potential for the inadvertent omission of monitoring points or important quality controls, missed key risks, or delayed staffing for central team roles and other unnamed and unexpected problems.

To the contrary, if we assume that the Project Manager has followed standing guidance regarding the development of the WBS and has engaged those who will be responsible for doing the work in the creation of this fundamental representation of the project's scope, then the WBS and WBS Dictionary provide a ready platform for framing the development of critical project support processes.

So when project managers comment they would prefer working through and developing project processes in an unstructured way, we counter by asking how they can be sure all aspects of the project's scope are addressed by the processes they've developed if they haven't begun with a breakdown of the project's scope—the WBS.

As the WBS decomposes the project's scope, it illuminates natural groupings of work for the Project Manager as well as the team. These work groupings lend themselves well to the analysis and construction required for developing support processes for the project. Using the House metaphor in Exhibit 6.1, we will apply the WBS to Quality, Risk and Resource Planning.

THE WHOLE IS NOT GREATER THAN THE SUM OF ITS PARTS—IT EQUALS PRECISELY 100% OF THE SUM OF ITS PARTS

Let's first look at the WBS as a whole. We'll break it up into segments to illustrate specific items later. When we look at this breakdown of the work, we immediately see there are some important differentiations among the work elements described by the WBS. If you can imagine the work actually being performed, you might envision that many specifically skilled technicians and laborers are required to complete the work

1 House Project
- 1.1 Primary Structure
  - 1.1.1 Foundation Development
    - 1.1.1.1 Layout – Topography
    - 1.1.1.2 Excavation
    - 1.1.1.3 ConcretePour
  - 1.1.2 Exterior Wall Development
  - 1.1.3 Roof Development
- 1.2 Electrical Infrastructure
- 1.3 Plumbing Infrastructure
- 1.4 Inside Wall Development: Rough Finish

Exhibit 6.1 House Example WBS

described by these simple entries. Sure, there may be a few individuals who have enough skill in each of the areas defined by the WBS to complete all the work themselves, but we would like to have you think of the more common approach, where today's home construction workers are more specialized and know their specific craft well. Home building has become another domain of skills specialization.

If we consider this to be the case, then it becomes abundantly clear from the decomposition of the work that the skills required for some of the WBS elements are separate and distinct from the skills required for others. In fact, workers you might identify as "critically needed" for some of the WBS elements would be entirely without value for others. For instance, you wouldn't want to have a certified electrician working on the plumbing infrastructure unless this person had some sort of certification in plumbing as well. They are quite clearly separate and distinct skills.

Another very important aspect we see as we look at the WBS as a whole is the fact that each of the sub-elements of the WBS is a complete component of work on its own, yet it must be able to be integrated and must come together seamlessly to produce the final completed product, the house. If these components do not fit together well, in all likelihood the purchaser will not accept the finished product.

So what can we surmise from this brief look at the WBS? First, we can use the WBS to "envision" or perceive how the work will be staffed with specific skills. Beyond that we can easily see there will be things that should be considered about how the various separate components must be developed in order for them to be integrated and functioning together as a whole. And thinking forward through the development process, we can imagine the testing and verification steps that will have to occur to ensure they do meet the applicable quality specifications and criteria defining "fitness for use."

Now let's discuss how we would use this WBS to develop processes for Quality, Risk and Resource Management specifically tailored to support this project. Here, we'll take a few of the WBS elements separately.

When we look at the level 3 and level 4 WBS (see Exhibit 6.2) elements that make up the Foundation Development components, we immediately see that there are definitions of work in this parent-child relationship that will require a few different skill sets. The work will also demand careful timing and a considerable degree of oversight.

The components we are discussing are shown here:

1 Foundation Development
  1.1 Layout – Topography
  1.2 Excavation
  1.3 Concrete Pour

Exhibit 6.2 House Example WBS Component

This is where most home construction projects begin. And given the fact that it represents the foundation for the remaining structure of the house, the quality of this component will have a great influence on the quality of the completed, finished home.

EXAMINING PROCESS CONSIDERATIONS

When considering risk and resourcing for the project, the Project Manager (or building contractor) for the house may be anticipating that each of these WBS elements will be subcontracted to different firms who specialize in the specific tasks described by the WBS elements. Here, the Project Manager would be balancing the staffing/specialization against the need to subcontract the work. To reduce risk and to improve overall quality, it is quite common in home building to find that construction organizations align with specific subcontractors who specialize in certain aspects of construction and will perform the same tasks on a series of home projects. Often subcontracts for this specific work within the WBS are set weeks, months and even years in advance. In this way, the contractor and subcontractor(s) have full knowledge of the anticipated work being performed (represented by the specific WBS elements describing the work) as subcontracted deliverables, and can write contracts articulating the relationship as well as the timing of the work. The WBS makes it easy to separate these deliverables into subcontracts, estimating and anticipating individual component costs within each of the WBS elements. This also enables oversight and management of the work as independent deliverables contributing to a cohesive whole, accounting for the cost of each child WBS element (deliverable) as a component of the parent, whether it is performed as a subcontracted set of deliveries or not. In this way, the Project Manager can draw a direct relationship between the WBS elements and Work Packages defined and articulated within the WBS, and the cost elements for each of those WBS elements and Work Packages represented in the model described by the Cost Breakdown Structure for the project and cost management tools used to monitor and manage the financial aspects of the project.

For example there may be a surveying firm familiar with the geography in the area in which the home is being built who may also have experience with the applicable laws, codes and regulations that must be considered when placing a home on a particular property. It would be important to subcontract the WBS element 1.1.1.1 Layout-Topography to such a firm to be sure that appropriate measures were taken to layout the foundation so that it falls within the boundaries and constraints defined by those laws, codes and regulations. Subcontracting this work solves the resourcing problem and allows the tracking of the cost for this work to be contained within this specific WBS element. Doing so, however, presents its own risks. Often firms that are familiar and experienced in performing specialized work are in high demand. Scheduling them to perform the work on specific dates can be a challenging affair. Schedules are frequently filled with previously contracted work: weather and unforeseen circumstances can impact the availability of the work crew, and due to the nature of the work, can impact the schedule for the remainder of the project.

The same would hold true for a subcontract organization that would be responsible for digging the foundation, WBS element 1.1.1.2, or the contracted team that would pour the foundation, WBS element 1.1.1.3. Each of these components can easily be seen as subcontracted parts of the Foundation Development components for the house, and at the same time can illustrate the risks inherent in subcontracting performance as well as accountability to third parties. While one aspect may be resolved, other risks are presented.

Here we can point to quality, risk and resource considerations associated with just these four WBS elements, while at the same time we can see how these elements present risk and potential impact to other elements in the project. While the prime contractor may subcontract some of the work to outside responsible parties, ultimately the prime contractor is accountable for all the work being performed according to specification, executed within budget and delivered according to the pre-determined schedule. Unfortunately, while the prime contractor may impose incentives on the subcontract organizations to ensure their performance and delivery, he or she cannot control their performance entirely. Events beyond the control of either party can impact delivery.

Additionally, it is clear that these three significant components of the WBS are interrelated. The layout and topography must be completed, and must conform to the applicable specifications relating to environment, lot placement, access, utilities and the like before a single shovel of dirt is taken from the ground. The concrete required for the foundation must be planned and ordered before the foundation has been started, or the prime contractor runs the risk of the concrete not being available when it is needed. If there is a delay to either the layout or excavation WBS elements, however, there is a high probability the concrete will be onsite before it is actually needed. With nowhere to pour it on arrival, the concrete would need to go back to the source. Rescheduling the availability of that concrete could present an additional set of challenges, risks and unanticipated cost to the Project Manager (contractor).

Quality Management processes relating to these same WBS elements are equally designed to meet the specific needs of the work being performed. The WBS enables the Project Manager to perceive the unique quality assessment, monitoring and controlling attributes associated with individual WBS elements themselves. For example, the quality assurance and quality control processes associated with layout and topography would be very different from the quality processes that would apply to the concrete poured into the foundation. Accuracy and conformance to codes or laws would apply to WBS element 1.1.1.1 (Layout-Topography), while plumb, level, square, drying time, strength, fit and finish would be appropriate for 1.1.1.2 (Excavation) and 1.1.1.3 (Concrete Pour).

This is, as you can imagine, a delicately scheduled and coordinated set of work. The WBS enables the project manager and the remainder of the project team (contractors, subcontractors) as well as stakeholders (the buyer, the mortgage holder, the subdivision owner) to group resource assignments (possibly through assignment of specific WBS elements to subcontract organizations) risks and quality management attributes. The WBS additionally enables the Project Manager to see the interrelationships between WBS elements and informs decision-making to ensure that the integration aspects of each of the individual WBS elements is addressed and managed.

While the WBS performs a vital role in the design and development of these processes, it also serves as a foundation for establishing entrance and exit criteria for the various stages of the processes themselves.

For example, *entrance* criteria for WBS element 1.1.1.3 (Concrete Pour) imply a set of *exit* criteria for WBS element 1.1.1.2 and might be listed as a set of criteria that includes:

- Foundation excavation (WBS element 1.1.1.2) is complete.
  - Floor is level and square (deviation tolerance +/– $\frac{1}{10}$ of a foot. Elevation is finished floor minus 4 inches).
  - Excavation walls have proper slope away from the floor area (approximately 2 inch slope).
  - Excavated floor area is 3 feet past the footprint outline of the building (allowing for proper placement of forms).
  - Foundation soil, rock, clay, etc. are dry (no visible standing water).
  - Forms are placed and secured.

Now that we have described important interrelationships between the WBS and a few of the key processes that are derived from it to support the project, we must examine the most critical aspect of project leadership, maintenance and control—Project Communications Management.

COMMUNICATIONS PLANNING USING THE WBS AS A FOUNDATION

Early in this chapter we mentioned that we would reserve the discussion of Communications Planning to later in the chapter. We've now arrived at the point we wanted to reach before beginning the discussion.

First and foremost, the WBS is itself a communications tool. The WBS articulates the scope of the project by decomposing it into clear, understandable component parts of the whole, each providing insight into specific deliverables that together make up the final end product, service, or result. This then becomes a representation of the project that can be shared between parties and discussed, revised, negotiated and approved. It can function as the primary representation of the agreements between the purchaser and provider, unambiguously describing what will be part of the project effort and what will not. When changes are proposed, the WBS can be the starting point for the discussions about cost, schedule and feature/functionality. We will discuss this in more detail in Chapter 8. On its own, the WBS performs an invaluable role as the historic (team memory) representation of the agreed-upon project outcome.

Beyond this, and regardless of the process used for managing Quality, Risk and Resourcing, the WBS should be consulted for development of Communications processes to support your projects. Why, you may ask. To answer that question, consider this: While many projects in an organization may be similar, no two can be exactly alike. Circumstances regarding specific components may be different, individuals who participated in one project may not be available for another, the customers or stakeholders may be different, and so on. The list could continue quite a bit longer. Whatever the driver, projects, by their very nature differ from one another. They have unique start and end points and are separate efforts precisely because they are different. If two projects were being conducted to produce precisely the same outcome at precisely the same time and in precisely the same space they would not exist as two distinct and unique efforts. They would be considered one project. Either that or some aspect of one of them would be different causing them to be separate project efforts. Remember, a project is "a temporary endeavor undertaken to create a unique *product, service or result*" (*PMBOK® Guide*—Third Edition, p. 368).

Because each project is different, each project will have different communications requirements. Again, these could be slight or small differences, but they will be different nonetheless. When this is the case, the Communications Plan and *communications needs* for the project should be examined carefully.

At first blush the Project Manager might assume that every member of the project "community" needs the same information presented in the same manner. To enable effective project communications, however, the

Project Manager discovers during the evolution of the project that each member group (the team, the sponsors, and the stakeholders) requires key information about the project presented in different ways. The WBS aids the Project Manager in tailoring project communications to meet the needs of specific groups and reveals the boundaries of interest for each group. Most importantly, the Communications Plan should describe how the project's information will be delivered to various project stakeholder groups in the method and mode *they* believe is most beneficial to them. Project communications are most effective when they are designed to meet the needs of the *receiver* rather than the sender.

Just as the WBS provides insight into Risk, Resourcing and Quality attributes associated with each unique deliverable, the Project Manager can use the WBS to construct an associated Communications Matrix to work in conjunction with the WBS.

DEVELOPING THE COMMUNICATIONS PLAN

The primary objective of the **Communications Plan** is to precisely match the type of project communication to the needs of the receivers of the communication, while making use of both a primary and backup or secondary methods for communicating a particular set of information. For example, where a face-to-face meeting may be the primary communication *method* for a particular event, one secondary *method* for communicating the same information could be through distribution of Meeting Minutes with an associated Action Register, sent through e-mail.

As the Communications Plan (and the conceptual communications platform or structure on which it relies) evolves, an effort should be made to utilize a primary and secondary *mode* for certain, if not all communications. To explain, at the current time the most common *mode* (or medium) used for distributing meeting minutes, action registers, Project Schedules and the like is direct e-mail, utilizing ad-hoc or stored distribution lists that the Project Manager maintains. In this *mode*, project team leaders *push* (send, without waiting for a request) information to project stakeholders, and as a result, participants and stakeholders receive what the Project Manager has determined would be important to them. This mode could, just as effectively, be complemented by, and in some cases entirely replaced, by an alternative method or mode, where project stakeholders access a Web site, an electronic data store or shared computing environment and *pull*, (select, read, print or view) only the specific project information they wish to see.

While today the primary *mode* for distributing Project Schedule information might be *push* to stakeholders and team members, a secondary, alternate *mode* could be the stored electronic version of the schedule, from which team members and stakeholders *pull* the information from a shared server or storage space. At some future time, the Project Manager could find that the preferred primary and secondary modes for project schedule information could be reversed.

Many additional modes of communication exist for information sharing. Face-to-face meetings, where the participants convene in a single room can also be "approximated" by using conference calls, video conferences, document-sharing facilities and/or *webinars*, where many people view and share a document while the meeting leader describes information in streamed video, projected slides or written documents.

THE COMMUNICATIONS MATRIX

To develop an effective Communications Plan, Project Managers may want to consider developing a **Communications Matrix**, where each individual associated with the project is listed, from stakeholders to developers, from customer representatives to engineering, construction, development, and support organizations. These lists can be directly associated to the various elements of the WBS, where specific information relating to individual deliverables can be summarized and presented as groupings of information related directly to groupings of work.

The Communications Matrix should include those who will directly conduct and perform the activities within the project or program, as well as those who will be affected by those activities. Here, the WBS Dictionary becomes increasingly valuable. The WBS Dictionary explains each WBS element in greater detail and can include individual roles, specific outputs, deliverables, and accountability, clarifying the information needs of those associated with the element. Specific roles for each individual, group, or team can also be defined in the Communication Matrix to help clarify the type of communications *channel* they might prefer, enabling the Project Manager to plan and design the most appropriate methods and modes for communicating with them.

A key element of the completed Communications Matrix will be the list of participants and stakeholders showing the most appropriate method and mode for communicating with them. A natural complement to the Communication Matrix would be a meeting and documented communications schedule that would outline the frequency, duration, and agenda for regularly scheduled meetings as well as the expected attendee list. The Communications Matrix might also outline the mode(s) used for delivery of hardcopy reports and publications to each stakeholder group and could include enough detail to address the communications needs of all members of the group.

When complete, this Communications Matrix becomes the cornerstone of the Communications Plan and can be updated along with the remainder of the project's documentation to ensure a direct, clear and concise body of information is communicated to those who need it, when they need it and how they wish to receive it. Again, the objective of the Communications Matrix is to provide a guide for delivering specific information to those who need it, utilizing methods and modes tailored to meet *their* needs. Table 6.1 (see page 104), an example from the New Mexico Department of Information shows how the Communications Matrix is developed and utilized.

THE HIERARCHY OF INFORMATION

In every project there are multiple channels for communicating information. By the very nature of the project structure, information generated at some levels of the project likely won't be important to participants and stakeholders at others. For example, meeting minutes and notes from daily or weekly working team meetings are rarely of use to project sponsors and corporate executives. To understand the activities of particular projects, key stakeholders at the executive level frequently require project summaries, single-page reviews, progress and highlight reports showing project milestones and their status, noting issues, caution or jeopardy. Just as meeting minutes would be of little value to the senior leaders of the project, working team members would probably find summary reports too broad and general to be of any use to them. An example of an information hierarchy is illustrated in Table 6.2 (see page 107).

Project Managers might find it useful to build a similar hierarchy based on the various elements of the WBS. It may be too cumbersome for the Project Manager to design a unique hierarchy for each of the individual WBS elements.But it may be quite useful to prepare a view of the information that would be relevant for each of the levels of the WBS.

Table 6.1 Communications Matrix

Name/ Nature of Communication	From	To	Content Provided By	Type (Man/ Mktg/Info)	Frequency	Format Used	Delivery Media	Comments
Sponsors								
Urgent Issues	Program Manager, Program Director	Executive Sponsor, Program Sponsor	Program Manager, Project Managers, External Stake-holders		As needed		E-mail	The Program Manager will collect this issue and add an entry in the Issues Log.
Issues Updates/ Resolutions	Executive Sponsor, Program Sponsor	Program Director, Program Manager	Executive Sponsor, Program Sponsor		As needed		Verbal updates, E-mail, Memos	The Program Manager will update the Issue and associated Log.
Status Report	Program Manager	Program Director	Program Manager, Project Managers	Mandatory	Monthly	Status Report form	E-mail or Shared Storage	The Program Manager will pull information gathered from the program status reports.

Special Presentation or Meetings for Updating Executives	Program Manager	Executive Team	Program Manager, Program Director	Informational	As needed	To be determined, based on requirements	Meeting	
Stakeholders								
New Issues/ Action Items	Stake-holders	Program Manager	Stake-holders		Bi-Weekly	Discussions during bi-weekly stakeholders meeting	Issues/ Action Items section of meeting minutes	The scribe will capture issues/action items and maintain a running log through the meeting minutes document. The scribe will also submit these issues/action items via the Program Manager database.

(continues)

Table 6.1 (*Continued*)

Name/ Nature of Communication	From	To	Content Provided By	Type (Man/ Mktg/Info)	Frequency	Format Used	Delivery Media	Comments
Issues/Action Items Status/ Updates/ Resolutions	Program Manager	Stake-holders	Program Manager, Project Managers, Team Members		Bi-Weekly	Program Management update during stakeholders meeting.	Stake-holders Meeting	The program manager will review the open issues/action items with the project teams and provide update, capturing all within the Program Manager database.
Urgent Information Impacting-Team and External (I/S) Stakeholders	Program Director	Team & External Stake-holders	Program Manager, Program Director, Project Managers		As needed	TBD	E-Mail/ Voice Mail, as appropriate	As critical information such as newly developed issues arise, a communication will be distributed to ensure immediate knowledge transfer.

Source: New Mexico Department of Information Technology. "Project Communication Matrix."

Table 6.2 Information Hierarchy Example

Level	Information Type
Senior Stakeholders (CIO, Project Sponsors, Customer Sr. Mgmt., Vendor Sr. Mgmt, Program and Project Leaders, Steering Committee, etc.)	Project Dashboard Milestone Charts Project Summary: Last 30 Days, Next 30 Days, Key Issues Budget Report Earned Value Report Risk Summary Issues needing executive intervention
Functional Directors, Matrix Organization Leaders, Project Managers, Account Managers, etc.	Project Schedule Summary and Milestone Chart Progress Chart Stoplight Chart Meeting Summaries Issue Logs Deliverable Schedules Specific Risk Impact Analysis Problem Notification (Escalation)
Component Managers, Operations Managers, Support Organization Managers, Project Managers	Detailed Project Schedule Detailed Risk Summary Meeting Minutes Dependency Matrix Action Registers
Etc.	Etc.

THE MEETING MATRIX

To be sure we remain on the same page, we must say one thing right here—meetings are communications too! Now that we've clarified that issue, let's move along.

Just as the **Hierarchy of Information** outlines the type of communications needed and shared with project team members, the stakeholder community and sponsors, the **Meeting Matrix** clarifies who the natural participants will be for each type of meeting, and indicates how frequently these meetings are held. The meeting matrix can include a great deal of information and again reflects the scheduling needs of the participants. The WBS describes the work along with the Organizational Breakdown Structure (OBS) that depicts the organizational relationship between and among the various project participants. The Meeting Matrix attempts to set out a schedule for each key project meeting type and includes enough detail to explain the intended content of the planned meetings, while helping to communicate the appropriateness of the meetings for the planned attendees.

To explain, we'll use an example. Project working level meetings occur very frequently, in fact, these are often held daily. While a senior leader or stakeholder may visit working-level meetings on occasion to observe or share key information, more than likely their schedules do not allow for many visits. Moreover, the detailed discussions that occur in these meetings are well beyond the scope of many leaders' interests or usefulness. In contrast, a senior-level discussion of the strategic positioning of the current project initiative might be an interesting event for a developer or engineer, but doesn't have a great deal of direct, relevant value to that individual. In the same way the Hierarchy of Information links the content to be shared with the needs of the individual or group, the Meeting Matrix outlines the participant list, the schedule and purpose for specific project pre-scheduled meetings. An example of a Meeting Matrix is found in Table 6.3.

Also, just as the Hierarchy of Information might be useful to the Project Manager for anticipating the informational needs of the project team and stakeholders, the Meeting Matrix can provide an easy explanation of key project events for these same groups and provides insight into upcoming events.

Table 6.3 Meeting Matrix Example

Level	Meeting Type	Frequency
Senior Stakeholders (CIO, Project Sponsors, Customer Sr. Mgmt., Vendor Sr. Mgmt, Program and Project Leaders, Steering Committee, etc.)	Stakeholder Review Steering Committee	Quarterly Monthly
Functional Directors, Matrix Organization Leaders, Project Managers, Account Managers, etc.	Project Leadership Team (Board Of Directors) Architectural Review Board Budget / Finance Team	Bi-Weekly Monthly Weekly
Component Managers, Operations Managers, Support Organization Managers, Project Managers	Direct Report Team Meetings Cross-Organizational Team meetings Project Team Meetings	Weekly Bi-Weekly Weekly
Etc.	Etc.	

CHAPTER SUMMARY

With the WBS and WBS Dictionary as the foundation for process development, the individual nuances of each process can be balanced against the content description for each deliverable. Whether the subject is Quality, Risk, Resource Management or Communications, the Project Manager is guided by the detail of the WBS and WBS dictionary. If stable, repeatable, processes do not exist in the project you are facing, the WBS can provide the bread crumbs you need to find your way out of the forest (and create the process).

As we have discussed in earlier chapters, one of the Core Characteristics of a quality WBS is that it is created by those doing the work. When team members are directly involved in creating the WBS, there is a high likelihood they will assure their work is represented within it. In fact, project team members who take part in these activities often insist their work is represented in the WBS, and will not allow the WBS to progress toward completion without the appropriate acknowledgment of their work. Rather than taking a passive approach to this, team members who participate in building the WBS more frequently will identify missing and omitted work elements.

By contrast, when project team members are presented with project elements (such as the WBS, Project Schedule or Status Report) that have been produced without their involvement, they often ignore or reject them.

REFERENCES

Covey, Steven R. 1989. *7 Habits of Highly Effective People.* New York: Simon and Schuster.

New Mexico Department of Information

Project Management Institute. 2004. *A Guide to the Project Management Body of Knowledge (PMBOK® Guide)—Third Edition.* Newtown Square, PA: Project Management Institute.

CHAPTER QUESTIONS

1. One sure path to success for the project manager is to apply tried and tested processes.
   a. True
   b. False

2. Which of the following can the WBS be utilized for? (Select all that apply.)
   a. Envisioning work to be staffed
   b. Understanding the integration of components
   c. Imagining testing and verification steps
   d. a and b only

3. Utilizing standard company processes for staffing individual project tasks is not beneficial to the project manager in organizations where processes are known and stable.
   a. True
   b. False

4. The ______________ serves as the foundation for establishing entrance and exit criteria for the various stages of the project.
   a. Work Breakdown Structure
   b. Project Schedule
   c. Project Charter
   d. Scope Statement

5. The WBS can be considered a ______________. (Select all that apply.)
   a. Communications tool
   b. Scoping tool
   c. Planning tool
   d. Measuring tool
   e. a, b, and d

Chapter 7

The WBS as a Starting Point for Schedule Development

CHAPTER OVERVIEW

We are presenting some critically important concepts in this chapter. As project management practitioners, we wrestle with these issues and concepts daily, and as authors and presenters we are frequently asked to comment on or explain our approach to one or more of these concepts. There is quite a bit of new information in this chapter as well. With this in mind, we want to highlight the important concepts sufficiently for you. This chapter includes the following:

- Demystifying the Transition from WBS to Project Schedule
- Applying Scheduling Concepts (Putting These Concepts to Work)
- Representing Scope Sequencing and Dependency
- Creating a High-Level Scope Sequence Representation
- The Concept of Inclusion
- The Scope Relationship Diagram
- Creating a Scope Dependency Plan from the WBS Element List and Scope Relationship Diagram

As we have discussed earlier and have learned through our own experience, there isn't a single "right" answer to many of the questions about how to apply a particular Project Management concept, tool or technique. In this chapter we present information about approaches that have worked well for us and provide guidance for others—or so they tell us.

What is particularly interesting for us is that while colleagues describe their project management experience, they reveal that the issues they face frequently aren't related to understanding how to use and apply an individual tool, process, concept or technique; but rather concern the need to integrate two or more of these to achieve a desired outcome. Their comments often settle on the fact that the needed guidance about the "best" (or even the worst) approach for doing this in many cases doesn't exist.

Here is what we mean: The authors have been discussing, writing about, presenting and demonstrating methods for developing effective Work Breakdown Structures for many years. While we have been doing this, others in the Project Management industry have been delivering the same types of communication on a broad spectrum of subjects–from Acceptance Criteria to Workaround Planning–and every project management subject in between. What we typically do not discuss or describe in much detail, however, is how some of these key elements perform (best) together. Regarding WBS development, practitioners tell us that using some of the guidance available to them through existing writings—and by applying a little practice they can learn to develop effective Work Breakdown Structures. They understand and can produce what most of us would consider *high-quality* Work Breakdown Structures (please see earlier chapters about this aspect of WBS construction). Once they've scaled and accomplished that particular mountain, however, they tell us they are presented with another, even bigger challenge–making all of the work they've put into developing the WBS relevant. For example, though they may have developed truly "beautiful" Work Breakdown Structures for their projects, they find it difficult to apply the fully elaborated WBS to the development of the Project Schedule.

Moreover, once the project is underway, maintaining the integration of these two principal Project Management components or applying the WBS in a meaningful way to other aspects of the project seems to be a near impossibility.

We are presenting the material in this chapter to address these challenges, to illustrate how the WBS performs as the starting point for Project Schedule development and how the WBS takes an active role during the project's execution as a basis for project control and decision-making. We have included some interesting subjects that we will discuss in detail throughout this chapter:

- Beginning with the WBS
- Transitioning from the WBS to the Project Schedule

- The concept of *Inclusion* as a new dimension in the representation of scope relationships
- The *Scope Relationship Diagram*–a new scope representation that will help relate the project's scope to the Schedule, Deliverables and End Products (outcomes)

We will now address each of these issues individually. First, we'll discuss one of the most challenging issues, transitioning from the WBS to the Project Schedule.

DEMYSTIFYING THE TRANSITION FROM THE WBS TO THE PROJECT SCHEDULE

The frequent complaints we hear about the relevance of deliverable-oriented Work Breakdown Structures seem to be attributed to the absence of clear guidance about the methodology used to apply this scope definition to other project processes, tools and tasks. In particular, the lack of helpful information about the processes used to apply *deliverable-oriented* Work Breakdown Structures to project scheduling is seen as the primary obstacle project managers face when attempting to employ deliverable-oriented Work Breakdown Structures as a basis for scope management and schedule development. The difficulty they encounter making the logical association and transition from the WBS to Project Schedule drives their reluctance to adopt the practice. In fact, much of the available documentation (e.g., Pritchard 1998) for applying Work Breakdown Structures to project scheduling suggests the development of "action-oriented" or "process-oriented" Work Breakdown Structures to ease the transition from the WBS to Project Schedule.

To correct and counter this confusing instruction, we refer you to key guidance for project managers in the *PMBOK® Guide*—Third Edition, Chapter 6. This chapter, "Time Management," contains much of the information required to explain and resolve the deliverable-oriented WBS–to–Project Schedule transition challenge. In truth, the key concepts for the transition are somewhat obscured by other important concepts presented in the chapter. Specifically, we will not be discussing every element presented regarding Time Management such as Activity Resource Estimating and Activity Duration Estimating. However the core elements that show the linkage between the deliverable-oriented WBS and the Project Schedule are surely present. In the upcoming paragraphs, we have extracted the elements that explain this transition. Specifically, we are going to look carefully at Activity Definition (*PMBOK® Guide*—Third Edition, Chapter 6, Section 6.1); Activity Sequencing (Section 6.2) and Project Schedule Development (Section 6.5). These are the fundamental concepts required to simplify the process. There are three key concepts that explain the transition. We summarize these for you in the next paragraphs.

Activity Definition describes the inputs, tools, techniques and outputs necessary to create the listing of activities that will be performed to produce desired project outcomes. The Project Time Management Overview (*PMBOK® Guide*—Third Edition, Figure 6.1, p. 140) and the detail found in this section clearly show the Scope Statement, WBS and WBS Dictionary perform as inputs to the Activity Definition process. The WBS and WBS Dictionary also support the tools for development of the Activity List, Milestone List and remaining outputs of the process, including Decomposition, Rolling Wave Planning and others. To explain, let's examine our house metaphor a little more closely. Using Exhibit 7.1 we'll focus in on two of the level 4 WBS elements 1.1.1.1 (Layout–Topography) and 1.1.1.2 (Excavation).

In this case, the WBS elements 1.1.1.1 and 1.1.1.2 are considered Work Packages–the lowest level of the WBS. When transitioning from the WBS to the schedule, however, this WBS element would be decomposed further, into Tasks, Activities and Milestones. Using a decomposition process similar to the one used to develop the WBS initially, the team

- 1 House Project
  - 1.1 Primary Structure
    - 1.1.1 Foundation Development
      - 1.1.1.1 Layout–Topography
      - 1.1.1.2 Excavation
      - 1.1.1.3 Concrete Pour
    - 1.1.2 Exterior Wall Development
    - 1.1.3 Roof Development
  - 1.2 Electrical Infrastructure
  - 1.3 Plumbing Infrastructure
  - 1.4 Inside Wall Development: Rough Finish

Exhibit 7.1 WBS House Project Elements–An Illustration

1 Foundation Development (WBS Element)
- 1.1 Layout—Topography (WBS Element, Work Package)
  - 1.1.1 Duplicate Topographical Drawings
  - 1.1.2 Verify Contractor(s) Schedules
  - 1.1.3 Conduct Survey
  - 1.1.4 Survey Complete (Milestone)
  - 1.1.5 Mark Property
  - 1.1.6 Mark Foundation Boundaries
  - 1.1.7 Layout–Topography Complete (Milestone)
- 1.2 Excavation (WBS Element, Work Package)
  - 1.2.1 Clear Property
  - 1.2.2 Haul Debris
  - 1.2.3 Dig Foundation
  - 1.2.4 Foundation Excavation Complete (Milestone)

Exhibit 7.2 Further Decomposed Foundation Development Elements

would set about breaking this WBS element down into logical units of work (or tasks) and associated activities with significant milestones. Exhibit 7.2 is one possible (fictitious) example of the decomposition of WBS elements (Work Packages) 1.1.1.1 and 1.1.1.2.

We'll capture the essential elements for producing the list for you. In simplified form, it would appear as Input (WBS and WBS Dictionary) to Process (Decomposition) to Output (Activity/Milestone List). Graphically, it would look like this:

Input	Process	Output
WBS and WBS Dictionary →	Decomposition →	Activity/Milestone List

Activity Sequencing explains how the project's activities, milestones and approved changes are used as inputs to the activity sequencing process. The tools for developing the outputs are also described, including the Project Schedule Network Diagram, updated Activity and Milestone Lists and include various network diagramming techniques, such as Precedence Diagramming Method (PDM) and Arrow Diagramming Method (ADM). A simplified view would be:

Input		Process		Output
Activity / Milestone List	→	Network Diagramming	→	Project Schedule Network Diagram

Schedule Development describes how the two processes, Activity Definition and Activity Sequencing are used to produce the end objectives of the process–the Project Schedule, Schedule Model, Schedule Baseline and other related schedule components. In this section of the chapter, we explain how the outputs of the two preceding processes are incorporated as inputs to the scheduling tools and scheduling methodologies to produce the project schedule. Simplified, this can be illustrated as:

Input		Process		Output
Activities / Milestone List Network Diagram	→	Scheduling Method/tools	→	Project Schedule

Summarizing the information found in these sections, the core elements that enable the elaboration and development of the Project Schedule begin with the Scope Statement, WBS and WBS Dictionary.

- Starting with the WBS work packages, these inputs are taken through a decomposition process to produce the project's Activity and Milestone Lists.
- These in turn, are inputs to network diagramming which produces the Project Schedule Network Diagram and updated Activity/Milestone Lists. The Project Schedule Network Diagram details how (in what sequence) the project outcomes will be achieved.
- Finally, the Project Schedule Network Diagram and updated Activity / Milestone Lists are then used as inputs to the project scheduling tools and methodology to generate the Project Schedule. Illustrated

in simplified process-flow form as before, the entire process can be summarized as follows:

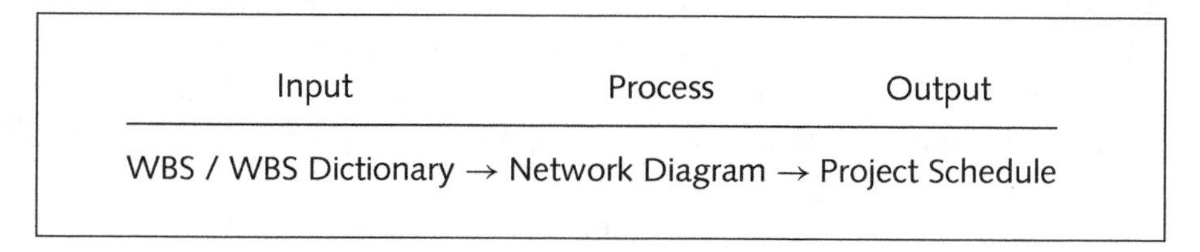

This simplified view in block diagram form is shown in Figure 7.1.

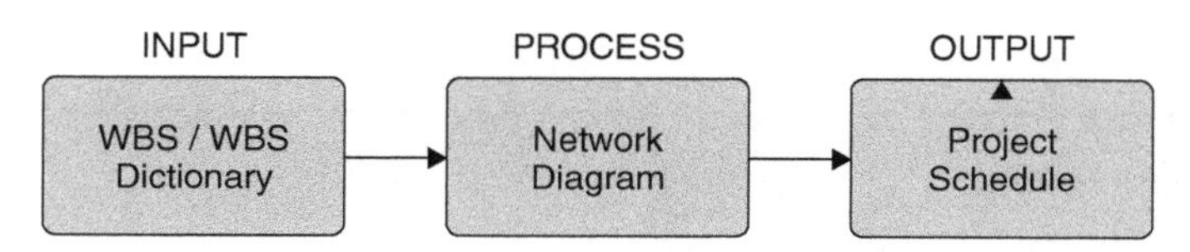

Figure 7.1 WBS to Project Schedule Transition.

PUTTING THESE CONCEPTS TO WORK

To illustrate how this process would be put into practice, we will continue with our simple house metaphor. To ensure we explain the construction of this WBS properly, we will presume for this discussion that the WBS elements listed in the outline are a few of the key scope components derived from an initial home building contract. Representing level 1, 2, 3 and 4 of the WBS, the high-level scope elements include the components of the primary structure, the foundation, exterior walls, roof, plumbing, electrical and interior walls. These eleven elements–without hierarchical structure—appear to the project manager (from the contractor) as follows:

- House Project
- Primary Structure
- Foundation Development
- Layout-Topography
- Excavation
- Concrete Pour
- Exterior Wall Development
- Roof Development
- Electrical Infrastructure
- Plumbing Infrastructure
- Inside Wall Development: Rough Finish

THE WBS IN HIERARCHICAL OUTLINE FORM

To organize this component list as it might be developed, the contractor working in conjunction with the project manager might organize the list of elements as they are depicted here. Even a novice would likely agree that these hierarchical relationships apply. For this example, we are presuming this is truly the correct representation.

In Exhibit 7.3, level 1 indicates the work called "House Project" which represents 100% of the work of the project. All other scope (WBS) elements associated with the project would be subordinate to the House Project element.

1 House Project
- 1.1 Primary Structure
  - 1.1.1 Foundation Development
    - 1.1.1.1 Layout–Topography
    - 1.1.1.2 Excavation
    - 1.1.1.3 Concrete Pour
  - 1.1.2 Exterior Wall Development
  - 1.1.3 Roof Development
- 1.2 Electrical Infrastructure
- 1.3 Plumbing Infrastructure
- 1.4 Inside Wall Development: Rough Finish

Exhibit 7.3 House Project WBS Elements–An Illustration

At level 2, there are four major components that make up the House Project:

- Primary Structure
- Electrical Infrastructure
- Plumbing Infrastructure
- Inside Wall Development

Level 3 shows the three key components of the Primary Structure:

- Foundation Development
- Exterior Wall Development
- Roof Development

And finally the Foundation Development is decomposed into three work elements that become level 4:

- Layout-Topography
- Excavation
- Concrete Pour

Granted, this is a highly simplified characterization of the work. It is used here, however, to help illustrate the Work Breakdown Structure's hierarchical concept, not necessarily the proper breakdown of all the work required to construct a home.

IDENTIFYING DEPENDENCIES BETWEEN SCOPE ELEMENTS

As we begin discussions of the following topics, the authors realize we are about to depart from traditional network/precedence diagramming and project activity sequencing guidelines. In the previous paragraphs we carefully described the processes that are followed to transition from the WBS to Project Schedule, and we pointed to the specific entries in the *PMBOK® Guide*—Third Edition where this process is explained. We believe this guidance is fundamental to the full understanding and use of the WBS and the proper approach to the transition from a conceptual view to a time-sequenced view of the project.

At the same time, we also recognize these closely guarded "rules" about precedence diagramming and scheduling may limit some of the additional value available to project managers through expanded use of the WBS. By introducing a few new concepts and processes for thinking through scope elements, we believe the project manager will greatly benefit from the WBS. Here we will more clearly document and detail interrelationships between scope (WBS) elements and begin to draw analogies between network diagramming techniques and scope–two concepts rarely referenced in the same sentence. With this in mind, we respectfully ask that you suspend disbelief once again, and join us in the discussion of some new approaches to scope definition.

REPRESENTING SCOPE SEQUENCE AND DEPENDENCY

Looking at the breakdown of the work previously described in Exhibit 7.3, contractors, project managers, homeowners and others would likely recognize that if this were the work to be completed, it would occur in a prescribed order, with some scope elements coming before and being completed before others begin. For example, it would be very helpful to build the foundation and walls before constructing the roof. Though it isn't mandatory to do it in this way, building the foundation first and then the walls; establishing this order would allow the roof to be constructed on top of the walls, where it will ultimately be completed and integrated to secure the structure. Certainly this is not the only approach to home construction, and the order can surely be modified to accelerate the building process, as is the case in modular home construction; but for this illustration, we will presume a traditional home construction project, and the order would be foundation, exterior walls, then roof.

Once the foundation, walls and roof are completed (and assuming additional details such as windows, doors and exterior finish are part of the work), the construction can move to the interior of the home. Here, it would make sense to complete the electrical and plumbing work before putting the interior wall material in place. As before, this order is not mandatory, but common practice would indicate the simplest, quickest and easiest approach would be to first complete the work that would be hidden by the interior walls, then apply the interior wall material. Again, for this example, we will use that convention.

CREATING A HIGH-LEVEL SCOPE SEQUENCE REPRESENTATION

With the previous discussion in mind, a project manager could begin developing a very high-level representation of the work described by the scope (WBS) using nothing more sophisticated than pencil and paper to illustrate the dependencies described. Beginning with the House Project

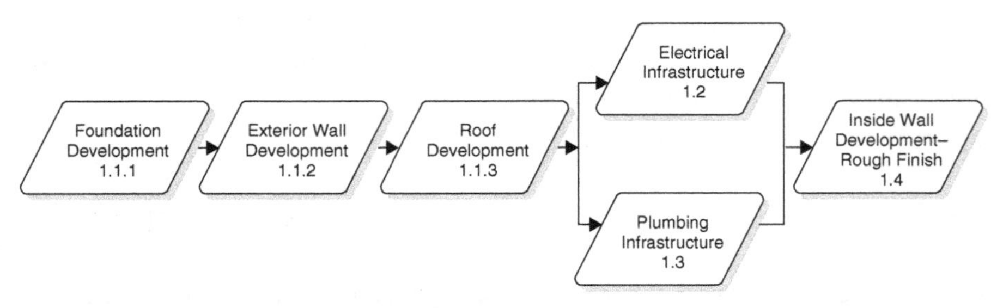

Figure 7.2 House Project High Level Scope Sequence.

element at level 1, and including all of the WBS elements required to show the implied dependency, one representation of the work might look like the set of interrelated elements found in Figure 7.2.

Figure 7.2 shows how the project manager would use a sequence representation–or an illustrated dependency map, to indicate that Foundation Development (with its Work Packages, Layout-Topography, Excavation and Concrete Pour) must complete before the Exterior Wall Development can begin, and that Roof Development depends on the completion of the Exterior Walls. Once the roof is complete, both the plumbing and electrical work can begin, but the Interior Walls would not start until the plumbing and electrical are complete. (In reality, the word "complete" here *could* mean "roughed-in" where wires and pipes are run to and from their destinations, but there are no fixtures attached to them.) It is important to note, the work elements shown here are not tasks or activities, but rather significant scope components that logically lead and follow one-another. Once these elements (Work Packages) are decomposed via the process described earlier, the resulting tasks, activities and milestones can be placed into the project scheduling tool.

THE CONCEPT OF INCLUSION

To further ease the transition from the deliverable-oriented WBS to Project Schedule, we can refine the central process to more clearly illustrate the relationships between scope elements before they are placed into a Scope Dependency Plan, which we will discuss later in the chapter.

In Figure 7.2, a high-level scope sequence was illustrated to show dependency between various WBS elements. In this example, each element is shown in linear fashion, using a two-dimensional format, with lines connecting the elements to show predecessor and successor dependencies. To produce this series of scope elements, the two dimensions at the core of the process are order (or "precedence") and dependency. While these two dimensions are critically important to development of a sequential illustration, in some cases they are not sufficient to enable the project manager to easily envision the *whole* project from the sequence.

Absent from this linear depiction of scope is the addition of a third dimension to complement precedence and dependency. To clarify, the concept or dimension of **Inclusion** can be added to the process to convert the linear, two-dimensional sequence into a graphic illustration that more accurately depicts how individual WBS elements relate to one-another, as parent and subordinate elements. The resulting illustration more accurately represents how the WBS elements appear in alternate representations, such as outlines, charts or WBS templates.

Inclusion as a dimension is used here to show which elements are "part of" larger scope elements as well as to clearly articulate which WBS elements are *not* "part of" the work of others. Said another way, some work depicted by a WBS is intended to be seen as being part of a higher-order work element, while other elements in the WBS are clearly not part of specific higher-order elements.

Using the example from the house project, we will take another look at the hierarchical outline for the work in Exhibit 7.4.

Describing this outline using the concept of inclusion it is easy to see that the WBS elements 1.1, 1.2, 1.3 and 1.4—Primary Structure, Electrical Infrastructure, Plumbing Infrastructure and Inside Wall Development - are all "part of" the House Project. They are integral to the completion of the project and are "included in" the work. By the same token, it is clear from the outline that the elements 1.1.1.1, 1.1.1.2 and 1.1.1.3 are all "part of" and "included in" the work that makes up the Foundation Development WBS element, element 1.1.1.

Our sequence diagram in Figure 7.2 shows the precedence and dependency between these scope elements, but does not clearly show which elements are actually part of the scope of other elements. In fact, if you examine Figure 7.2 carefully, you will notice that some of the elements

1 House Project
- 1.1 Primary Structure
  - 1.1.1 Foundation Development
    - 1.1.1.1 Layout–Topography
    - 1.1.1.2 Excavation
    - 1.1.1.3 Concrete Pour
  - 1.1.2 Exterior Wall Development
  - 1.1.3 Roof Development
- 1.2 Electrical Infrastructure
- 1.3 Plumbing Infrastructure
- 1.4 Inside Wall Development: Rough Finish

Exhibit 7.4 House Example WBS

have been left out of the diagram–for example, the level 1 WBS element House Project is not included. Additionally, the first level 2 element, Foundation Development is excluded, as are the three level 4 elements, Layout, Excavation and Concrete Pour. Why have they been excluded? Because including them in this drawing would be confusing and would disturb the illustration of the dependencies that are present. How would it be possible in Figure 7.2 to represent the level 1 or level 4 WBS elements without disturbing the logical flow of the dependencies between the relevant elements? In truth, it is nearly impossible to properly include those elements in this illustration. To correct this issue and explain, we will examine the Foundation Development elements closely.

In Figure 7.2 the Foundation Development elements at level 4, Layout-Topography, Excavation and Concrete Pour were excluded to reduce the confusion about the dependency between the level 3 elements, Foundation Development (1.1.1), Exterior Wall Development (1.1.2) and Roof Development (1.1.3). If we were to include them, however, they would also reflect their own natural or logical sequence. For instance, the layout of the foundation must precede any excavation–and the excavation must be complete before any concrete is poured. Considering the dependency between these elements, they could be shown as a series of scope elements executed in sequential fashion, under the "parent" element "Foundation Development" at level 3. This concept is shown, as an excerpt from the House Project, in Figure 7.3.

In this excerpt, it's difficult to clearly envision or understand the relationship between the parent and children WBS elements other than

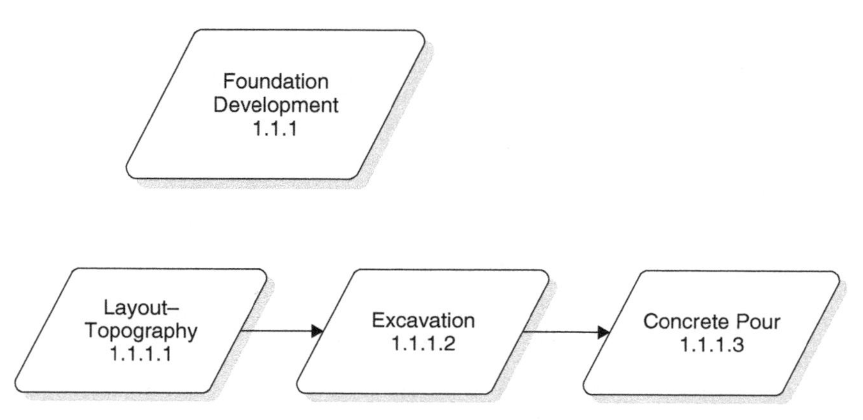

Figure 7.3 Foundation Development WBS elements from the House Project.

1 Foundation Development
  1.1 Layout–Topography
  1.2 Excavation
  1.3 Concrete Pour

Exhibit 7.5 Foundation Development Outline from House Project

the fact that we have told you the three elements at level 4 are children of the parent element Foundation Development–which is not accurately represented in Figure 7.3. If we were to link the parent, the Foundation Development would appear as simply another node in the sequence, when in actuality it isn't. In truth, the relationship between the Foundation Development element at level 3 and its children at level 4 is more clearly shown in the textual, outline form in Exhibit 7.5.

Here, it is easy to recognize the parent-child relationship between the level 3: Foundation Development WBS element and the level 4 elements, Layout–Topography, Excavation and Concrete Pour. Because of the indentation of the level 4 WBS elements under the parent element, this outline form communicates to us and clearly shows that Layout-Topography, Excavation and Concrete Pour are actually "part of" and "included in" the work that is called Foundation Development. Showing this in graphic format (see Figure 7.4) using an alternative view to represent this parent-child relationship may help somewhat, but does not fully capture the true relationship between the parent and child elements.

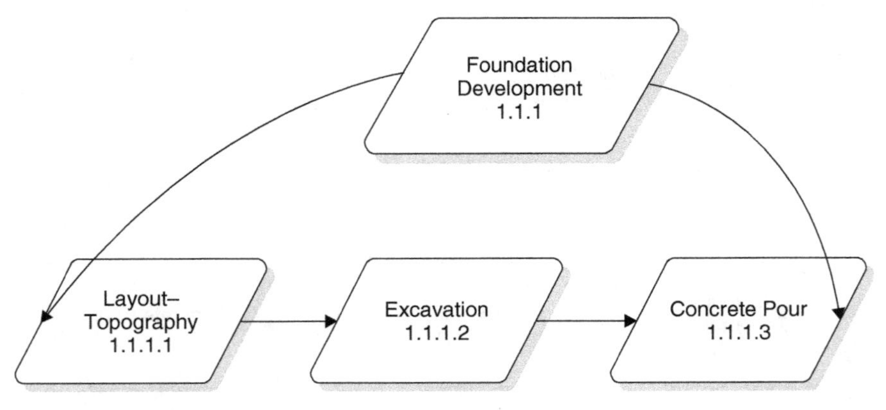

Figure 7.4 Alternate Foundation Development Graphic from House Project.

In Figure 7.4, it is difficult to determine the true relationship between the parent and child elements. Does "Foundation Development" come before or perhaps after the child elements? Of course, neither of those would be correct. Is Foundation Development above or below? Neither of those would be correct. Clearly, we need a better way to represent and communicate the relationship between these elements.

THE SCOPE RELATIONSHIP DIAGRAM

To solve and illustrate how these relationships actually occur, a **Scope Relationship Diagram** will be used to clearly show the relationships detailed in the outline version in Exhibit 7.5, as well as the order and precedence shown in Figure 7.3.

The resulting *Scope Relationship Diagram* reflects the added dimension of *Inclusion* representing these same WBS elements as follows in Figure 7.5.

Here, in this Scope Relationship Diagram representation, the Foundation Development WBS element—1.1.1 is larger and visually *includes* the lower level elements 1.1.1.1, 1.1.1.2 and 1.1.1.3.

With the addition of arrows to show the scope sequence described earlier, we are now able to illustrate how scope elements are planned within the concept of inclusion. In Figure 7.6 it is clear to see that the three elements at level 4 are executed in sequence "within" or as "part of" the scope of the parent element, Foundation Development.

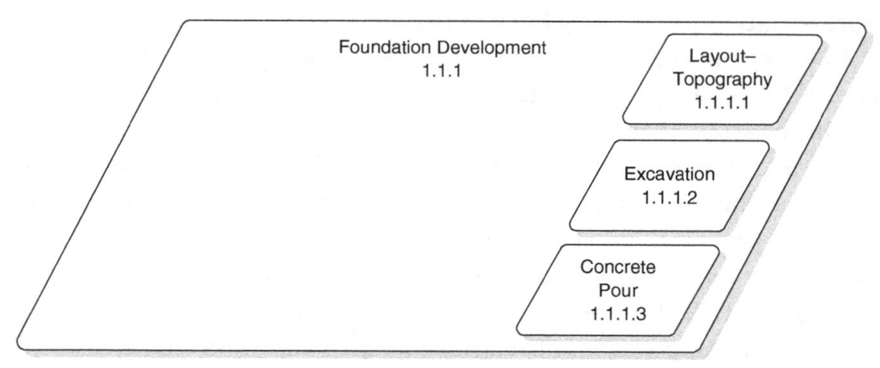

Figure 7.5 Scope Relationship Diagram from House Project Foundation Development Segment.

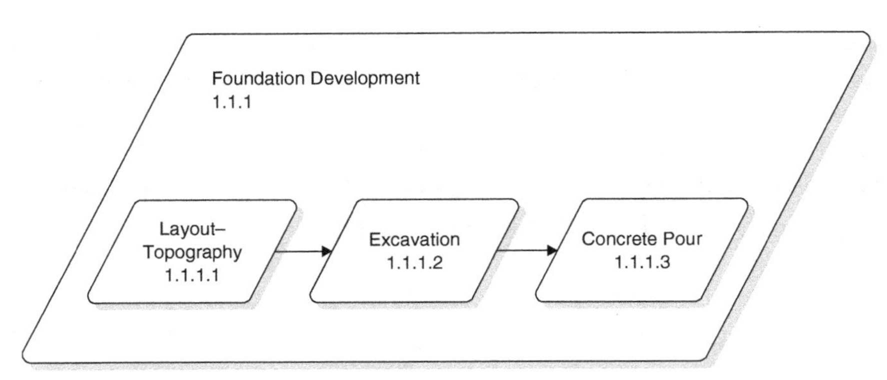

Figure 7.6 Scope Relationship Diagram from House Project–With Scope Sequence Foundation Development Segment.

Expanding this concept further to include all of the elements in the House Project, a Scope Relationship Diagram showing 100% (Core Characteristic) of the work defined in the outline presented in Exhibit 7.4 would produce the visual graphic illustrated in Figure 7.7 (see page 127).

With this illustration, demonstrating or describing which WBS elements are "part of" others is easy. The parent elements always include the child elements, and appear as nested representations of work within the Scope Relationship Diagram. Moreover, it is easy to recognize which WBS elements are both parent and child. Nesting the scope elements clarifies the true relationship between the elements, a representation that previously could be illustrated only in outline form. Finally, all the guidance we have shared with you about the development of "quality" Work Breakdown Structures remains intact with this representation. Core and Use-Related Characteristics are unchanged and apply to this representation just as they have in previous representations.

To take this concept further, while the Scope Relationship Diagram for the House Project enables the visualization of the work "included" within the scope of each parent WBS element, it also allows a more direct and straightforward transition from deliverable-oriented WBS to project schedule. This results from the additional clarity the Scope Relationship Diagram provides, as it represents the relationships between WBS elements graphically, showing how they interact within the entire scope of the project. Added benefits are also derived from this WBS representation. As decomposition is performed against the WBS elements in this Scope Relationship Diagram (the lowest level being Work Packages), the resulting tasks, activities and milestones can be easily grouped in the

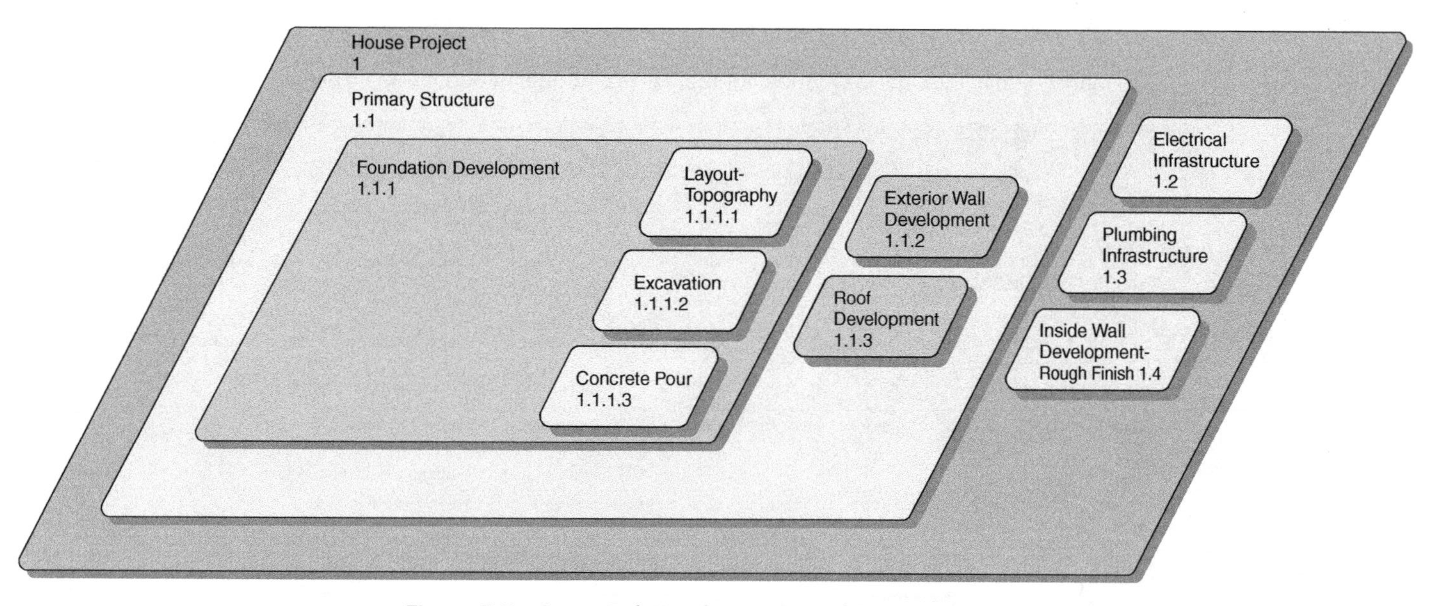

Figure 7.7 Scope Relationship Diagram for House Project.

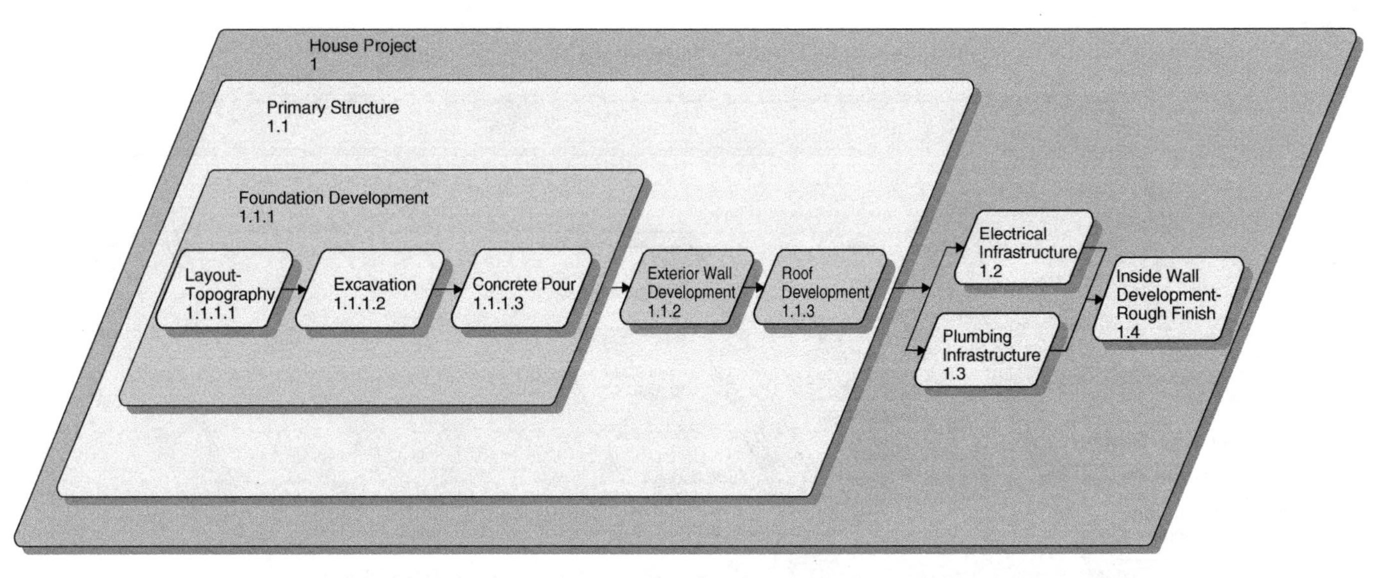

Figure 7.8 Scope Relationship Diagram for House Project–with Scope Sequence.

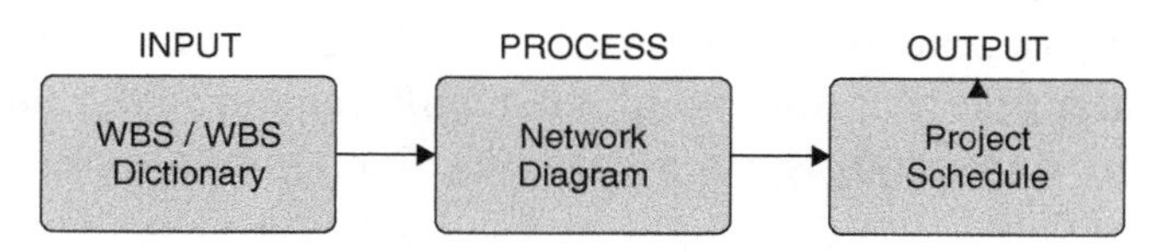

Figure 7.9 WBS to Project Schedule Transition.

same manner as the WBS. These will be input to the Project Schedule and will facilitate the grouping of work that will be monitored and controlled during the execution of the project.

Beyond the initial view in Figure 7.7, the various WBS elements can then be moved into a logical sequence. Dependency lines can be added to illustrate how the sequence of each of the scope elements within the project (parents and children) relate to and depend on one another. This reveals a logical representation of the sequence of the work to be performed. Using the Scope Relationship Diagram from Figure 7.7, adding the dependency lines would produce the logical sequence shown in Figure 7.8.

In this way, the project manager is able to use a step-wise process to create the linkage between the components of the deliverable-oriented WBS and the scope of the project, prior to further decomposition and development of the Project Schedule. Most importantly, representing the WBS in this way may *simplify* the transition from WBS to a Project Schedule we described at the beginning of the chapter. In the next paragraphs we illustrate how that is accomplished, but want to be sure you are able to clearly see these two methods as reliable ways to transition from the deliverable-oriented WBS to the Project Schedule. So to recap, a clear path can be drawn from deliverable-oriented WBS to Project Schedule, if that path is taken through a logical sequence of decomposition and network diagramming. This concept is represented in Figure 7.9, which is a repeat of the concepts we discussed at the beginning of the chapter.

CREATING A SCOPE DEPENDENCY PLAN

Beginning with the Scope Relationship Diagram represented in Figure 7.8, we will show the transition from that illustration to an initial text version of the same information, and we will refer to the end product as a **Scope Dependency Plan**. Figure 7.10 (see page 131) shows a Scope Dependency Plan as an excerpt from the project developed using a project scheduling tool readily available in the market today. In the illustration for this plan, you might recognize the output and immediately conclude this is a *project schedule* representation. It is not. We've anticipated that you might be thinking something doesn't sound quite right, something in this description doesn't fit together correctly —and it is true, at first blush, this would appear to be a contradiction of the rules and guidance we provided at the beginning of the chapter. Frankly, it's fitting to conclude this chapter with another, though small, departure from the norm. So bear with us; we'll explain.

In this illustration, we are simply representing another model as a starting point for the decomposition described at the beginning of the chapter and have substituted a new process for the center box in the process. The primary difference is that we've used a different methodology to arrive at the point of decomposition to tasks, activities and milestones. First, we've illustrated the WBS (scope) as a Scope Relationship Diagram. Next, we've demonstrated the *sequence* of the large scope components using the Scope Relationship Diagram reoriented to reflect the natural *dependency* between elements, and we've shown that dependency by adding connectors between the relevant WBS elements. Finally, we took the WBS elements (and their relationships) from the Scope Relationship Diagram and placed them in a tool that enables the linking of the WBS elements as they appear in the Scope Relationship Diagram with the Scope Sequence represented. What this allows is a quick transition from deliverable-oriented WBS to *Scope Dependency Plan* through the *Scope Relationship Diagram* (with Scope Sequence). The process can be summarized as we have at the start of the chapter. Here is a simple illustration of the process:

Input	Process	Output
WBS / WBS Dictionary →	Scope Relationship Diagram →	Scope Dependency Plan

In Figure 7.10 you will see the example of the Scope Dependency Plan that represents what we have been discussing throughout the book, the House Project. It includes the same eleven WBS elements we've shown and used repeatedly. Here, the leftmost column contains the line numbers for each of the WBS elements (and you've come to know them as parent

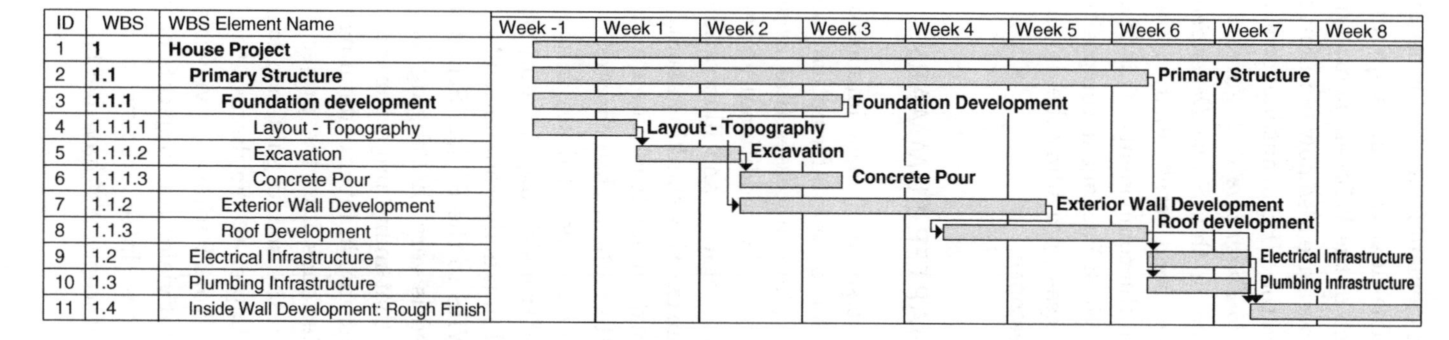

Figure 7.10 House Project High Level Scope Dependency Plan.

WBS elements, child WBS elements and the lowest level in the representation, Work Packages). The next columns show the WBS hierarchical numbering scheme for each of the WBS elements and the WBS element names from the project. These are also now familiar. In this example, the project manager referenced the WBS in outline form shown in Exhibit 7.4 and the *Scope Relationship Diagrams* in Figures 7.7 and 7.8 to create the dependencies described in this *Scope Dependency Plan*. The hierarchical relationships have been established in the tool to reflect the interrelationships between the components. This is also a reinforcement of the existing process for Work Package decomposition (the next step in the Project Scheduling process), providing the development of the initial Tasks, Activities and Milestones.

CHAPTER SUMMARY

In this chapter we present some of most important concepts we have included in this book. Some of the information found in this chapter is intended to clarify or explain processes that exist today, but remain illusive to many practitioners. We additionally present some very new concepts about WBS representation and illustration. The new concepts are a bit of a departure from existing thinking about WBS representations and decomposition, but are designed to reinforce and simplify the practice we have been discussing throughout each chapter.

First and foremost, we have included key information to clarify and ease the difficult process of transitioning from the representation of work represented and captured in a deliverable-oriented WBS. The *transition* we refer to is from the WBS to the Project Schedule, and it represents a process we have learned is very challenging for project managers. It is so challenging, in fact that many project managers simply choose to skip the WBS development step in favor of defining the work by entering tasks, activities and milestones into the project scheduling tool. For this discussion we rely on the sound guidance found in the *PMBOK® Guide*—Third Edition, Chapter 6. We discuss the stepwise process for the decomposition of WBS Work Packages, review Network Diagramming and discuss the process that yields the Project Schedule.

Following this discussion we introduce new concepts. The first of these is the idea that elements of scope (that have not been decomposed beyond the Work Package level) can be placed in sequence and represented as linked components of a project. Though we may know this intuitively, until now only tasks, activities and milestones have been represented in sequence developed as part of the project scheduling and network diagramming processes. In this section, we show how large groupings of work (or scope elements) can be sequenced, and are in fact executed in sequence during project performance. Representing this *Scope Sequence* is a relatively simple exercise. We illustrate these sequenced relationships as we would the sequence and dependencies that occur between tasks and activities in a project.

The next new element in this chapter is the concept of *Inclusion*. This is a scope development/scope illustration dimension that clarifies the relationships between and among elements. The concept of *Inclusion* allows us to communicate which scope elements are "part of" or "included in" other scope components. At the same time, this dimension clearly shows which scope elements are not part of other elements.

Finally, we introduce two new representations. The first is the *Scope Relationship Diagram*. This new way of illustrating the WBS (scope) allows the project manager to graphically demonstrate the relationships described through the previously described concepts of sequence and inclusion. Bringing all of these new concepts together into an integrated framework, the output is a representation of the high-level scope showing sequence and dependency of high-level scope components in the *Scope Dependency Plan*, the starting point for decomposition of the WBS Work Packages into tasks, activities and milestones.

The concepts discussed in this chapter are presented as a reference and resource for project managers, and as a means to a desired end—simplification of the application of the WBS, supporting its active, expanded use during project scheduling (Planning), Executing, Monitoring and Controlling and Closing.

REFERENCES

Pritchard, Carl. 1998. *Rational Unified Process*. Available at http://www.ts.mah.se/RUP/RationalUnifiedProcess/manuals/intro/im_diff.htm.

Project Management Institute. 2004. *A Guide to the Project Management Body of Knowledge (PMBOK® Guide)—Third Edition*. Newtown Square, PA: Project Management Institute.

CHAPTER QUESTIONS

1. Which term describes the inputs, tools, techniques, and outputs necessary to create the listing of activities that will be performed to produce desired project outcomes?
   a. Activity Definition
   b. Activity Sequencing
   c. Activity Estimation
   d. Work Breakdown Structure

2. What term explains how the project's activities, milestones, and approved changes are used as input to the activity sequencing process?
   a. Work Breakdown Structure
   b. Activity Definition
   c. Activity Estimation
   d. Activity Sequencing

3. Place the following deliverables in the proper sequential order of development by either filling in the blank table cells with the proper number or by reordering the deliverables

Order	Deliverable
	Network Diagram
	Project Schedule
	WBS / WBS Dictionary

4. Match each of the following elements to either a Work Breakdown Structure or a Project Schedule by filling in the table cell with the proper indicator—"Project Schedule" or "WBS."

Milestones	
Tasks	
Work Packages	
Activities	

5. Fill in the blanks with the appropriate words:
   Inclusion as a dimension is used to show which elements are ________________ larger scope elements, as well as clearly articulating which WBS elements are ________________ the work of others.

Chapter 8

The WBS in Action

CHAPTER OVERVIEW

The Planning portions of this book described how the Work Breakdown Structure is used as the basis for most of the project management Initiating and Planning processes and deliverables. This can be thought of as the setup discussions. As we now turn our attention to executing the project, including the monitoring and controlling aspects of project management, the WBS becomes the baseline, or foundation upon which execution is based. This is the point where the WBS transitions from a passive role to one of action. The WBS is now called upon as a foundation for decision-guidance. It is at this point where the rubber meets the project management road—the point where managers must manage and leaders must lead.

This chapter covers the following topics:

- Acquiring the Project Team
- Directing and Managing Project Execution and Integrated Change Management
- Performing Scope Management (which includes Change Management and the project management triple constraint)
- Reviewing the Relationship with Other Project Management Processes
- Performing Quality Assurance
- Performing Scope Verification

ACQUIRING THE PROJECT TEAM

As projects and programs begin, an important activity is acquiring the team—staffing. Without the appropriate resources, the project or program will not be executed nor completed. As we saw in the planning processes, the Work Breakdown Structure forms the basis for the Staffing Plan. As we move to and through project execution, the project team staff is added and released based on the needs of the project which are detailed in the Project's Staffing Plan. At all times, the WBS and WBS Dictionary are used to continually verify and validate that the appropriate resources are available and assigned to deliver the product, services, or results defined by the individual WBS work packages. Resources can come from a variety of sources—internal to the organization, external to the delivery organization or contracted externally from outside the company.

In many organizations, the WBS Dictionary is used to provide a detailed explanation of each project deliverable and define the boundaries as well as completion and acceptance criteria for the deliverables. Another use of the WBS Dictionary is to define the resources required to complete the deliverable, including the specific skills and competencies for those resources. This enables integration of the WBS Dictionary with a Responsibility Assignment Matrix (RAM), and provides many advantages. Among these is the ability to use the two key documents as a baseline reference during resource acquisition, management and resource releases.

Table 8.1 (see page 139) shows a portion of the WBS Dictionary for our example, the house. Due to space limitations, only the portion of the RAM integration (or RACI chart as it is sometimes called) with the WBS Dictionary is shown. In this example, the RAM columns depict the individual types of resources assigned to each of the work packages and how they are involved—Responsible, Accountable, Consulted, Informed (RACI). For some projects, specific resource names or even skill sets and competencies could be in the columns rather than just the types of resources. With this type of information defined in the WBS Dictionary, the Staffing Plan, WBS and WBS Dictionary become key reference documents for aligning the work to the available resources, managing variations to the plan and monitoring delivery.

Table 8.1 WBS Dictionary

WBS#	WBS Element	Responsibility Assignment Matrix					
		Architect	Prime Contractor	Builder	Electrician	Plumber	Inspector
1	House Project						
1.1	Primary Structure						
1.1.1	Foundation Development						
1.1.1.1	Layout—Topography	C	A	R			
1.1.1.2	Excavation		A	R			C
1.1.1.3	Concrete Pour		A	R			C
1.1.2	Exterior Wall Development		A	R			C
1.1.3	Roof Development		A	R			C
1.2	Electrical Infrastructure		A		R		C
1.3	Plumbing Infrastructure		A			R	C
1.4	Inside Wall Development: Rough Finish	C	A	R			C

R = Responsible *A = Accountable* *C = Consult* *I = Inform*

DIRECTING AND MANAGING PROJECT EXECUTION AND INTEGRATED CHANGE MANAGEMENT

During project execution, Work Breakdown Structures are used to continually verify and validate project scope against what has been agreed-to and baselined by the team. Beyond this, the WBS and WBS Dictionary form the basis for scope management, which is monitored and controlled via integrated change management. As the project or program progresses, variances occur and issues arise. As these variances and issues are evaluated and addressed, they may in turn lead to **Change Requests**—requests for modifications to the agreed-upon, baselined scope.

When Change Requests are raised and presented, they are thoroughly evaluated by key project team members to determine the impact the requests may have to the existing project scope, schedule and budget. To determine this impact, the size, complexity, timing and urgency are balanced against the existing schedule, available resource pool and expected spend for the project. The Work Breakdown Structure provides the baseline for the evaluation and estimation of these Change Requests. When the analysis and estimates are complete, they are presented to the project's sponsors and key stakeholders for determination of appropriate action. If the Change Requests are determined by project decision-makers to be mandatory, the budget, schedule and staffing plans will be modified to accommodate the changes. If, on the other hand, it is determined that one or more of the changes are not mandatory, it may be determined by the project stakeholders that the best course of action will be to stay on the current schedule, staffing plan and budget, and plan these changes for a later release.

Approval of Change Requests lead to a series of steps that must take place to ensure the project's scope, plan and budget remain current. These changes are managed through the Project's Change Management process. The steps include the following:

- Update of Scope Statement
- Update of Work Breakdown Structure
- Update of WBS Dictionary
- Update of all planning documents, including the Staffing Plan and Project Schedule
- Update of the project's budget

In short, the Work Breakdown Structure ties together all project and program sub-components and ensures that the interrelationships between them are managed so the project outcomes may be delivered. This process provides the foundation basis for the following:

- Clarifying Acceptance Criteria
- Development of Performance Criteria
- Transition planning for product, services or results and all documentation
- Quality Assessment and User Acceptance Testing
- End-user education and training
- Deliverable ownership transfer
- Completion of contractual obligations and contract closure
- Establishment of maintenance and support (post acceptance and transition to production)
- Explanation of warranty period

PERFORMING SCOPE MANAGEMENT

One of the most common uses of the Work Breakdown Structure is its use as the cornerstone for scope management and control. Process issues and variance, external decision-making, requirements changes and unexpected events (risks/issues) often lead to potential changes to expand or revise project scope. When these Change Requests arise, they must then be evaluated against the original project scope to determine the ultimate impact to scope, schedule and budget. Once again the baseline for this evaluation is the WBS and WBS Dictionary. As decisions are made to accept project changes, this in turn triggers updates to the project's Scope Statement, Work Breakdown Structure, WBS Dictionary, supporting Project Management plans, Project Staffing Plan, Project Schedule and Project Budget.

Another aspect of scope management is the use of the WBS to monitor supplier performance against planned deliverables. Here again the WBS and WBS Dictionary provide a baseline for measurements and monitoring. As work (depicted in specific WBS elements) is contracted and sub-contracted, the management of suppliers becomes another key aspect to the management of project scope. Just as the Responsibility

Assignment Matrix can be integrated into the WBS Dictionary, so too can supplier performance criteria for the WBS element deliverables to be cross-referenced in the WBS Dictionary.

A well-defined WBS with a detailed WBS Dictionary is the best defense against inadvertent scope creep for projects and programs. Combined with rigorous Change Management, this helps ensure that the project delivers only those products, services and results that are agreed upon, nothing more and nothing less.

SCOPE MANAGEMENT AND THE TRIPLE CONSTRAINT

One of the most fundamental aspects of project management is the framework of the **Triple Constraint**—scope, time and cost (see Figure 8.1). The triple constraint framework states that project quality is affected by the balancing of these three constraints. The relationship among these factors is such that impacting one factor will cause one or both of the other factors to be impacted as well. (*PMBOK® Guide*—Third Edition, p. 8).

Scope is one of the three factors that make up the triple constraint. As such, the management of scope, and the potential for changes to scope, will have an impact on time, cost or both. Given this, it is crucial that a balance be struck between the management of scope with the management of time and cost. All three factors use the Work Breakdown Structure as a starting point. Given this, the quality of the WBS and WBS Dictionary are crucial to the successful delivery of projects and programs.

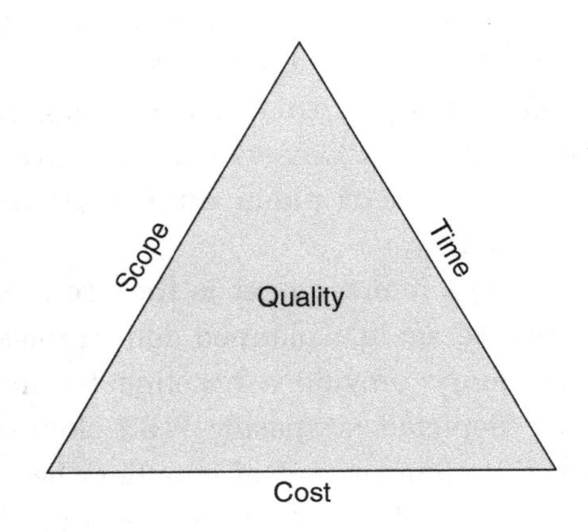

Figure 8.1 Triple Constraint.

As an example of how these three elements interact with one another in the triple constraint, let's refer to the house example. Let's say that part of the way through the project a Change Request is being evaluated to add another room and a deck to the house. This addition of another room and a deck increases the scope of the project. In evaluating this scope change, the following is determined:

- Some re-work of previously completed deliverables will be required (time and cost impact)
- Additional building materials will be required (time or cost impact)
- Additional time or additional resource will be required for the completion of the additional room and deck (time or cost impact)

While there most certainly will be other impacts, you can see from this simple analysis how a change to the project's scope impacts both time and cost.

Conversely, it is also true that changes to time or cost impact scope. How many times has a Project Sponsor come to the Project Manager and asked for certain deliverables to be completed ahead of schedule? In our experience, this is common. Generally speaking, moving the schedule in for some deliverables could be accomplished by removing other deliverables (scope reduction) or even by adding project resources to accomplish the given scope more quickly (cost increase).

These examples show how scope, time and cost (the project management Triple Constraint) are very much inter-related. As previously stated, the WBS and WBS Dictionary are critical to the management and control of all three constraints.

REVIEWING THE RELATIONSHIP WITH OTHER PROJECT MANAGEMENT PROCESSES

As is the case with Scope Management, the WBS and WBS Dictionary provide the foundation for decision making relative to other Project Management processes. With action and issue management, the WBS is often used to provide grounding for discussions on specific issues and action items. Doing this ensures alignment between the action or issues being discussed and the scope and objectives of the project or program. This is also true for risk management, financial management, earned value analysis as well as many other project management processes.

Using the WBS as the foundation for analysis and decision making ensures continued alignment with the project's scope, thereby helping to maintain the balance of time, cost and scope (the Triple Constraint).

Another benefit of aligning issues, action items, risks and the like around the Work Breakdown Structure is the ability to view WBS elements and work packages grouped as they are managed. This is especially useful in status reporting and project communications. By grouping these items, the Project Manager can provide another valuable viewpoint as well as further focus and clarity into the status and current operations of the project.

PERFORMING QUALITY ASSURANCE

The completed WBS, while under the control of change management, is the baseline document used to plan and perform Quality Assurance. Quality measurements and metrics are derived from the individual scope elements that are defined and explained in the WBS and WBS Dictionary. Without the WBS, measurement and metrics would be nearly impossible to define, let alone manage and execute.

As has been noted numerous times previously, the WBS provides the basis for the monitoring and controlling of the project or program. Process quality uses the WBS to establish key monitoring points (the WBS elements themselves) to indicate key deliverables and the statistics related to their performance. If defined properly, the WBS provides the baseline that allows for quality planning and assurance, and therefore leads to projects delivering to appropriate quality standards.

PERFORMING SCOPE VERIFICATION

During project execution, validation of the deliverables can be accomplished by referencing the deliverables as they have been described in the WBS and WBS Dictionary. Since the WBS and WBS Dictionary each describe project deliverables including acceptance and completion criteria, these then become the reference point for validation and acceptance of the completed deliverables. The WBS and WBS Dictionary often are used additionally as a baseline for monitoring and measuring "wants" and "needs" versus the agreed-upon project scope. This ensures that the project does not attempt to deliver outcomes that are not included in the requirements. The WBS and WBS Dictionary help ensure the project team does not attempt to deliver outcomes or quality that exceed the boundaries of the requirements while they also help contain and control scope creep.

The WBS and WBS Dictionary help support communications between the project manager, project team, sponsor(s) and stakeholders regarding the content and completion criteria for project deliverables. Without first developing the WBS, frequently the criteria for deliverable acceptance and completion are ill-defined, leading to misunderstanding and disagreement about the completion of specific project outcomes.

As work proceeds on the project, the WBS can be used as a checklist to determine what deliverables have and have not been completed or accepted. When communicated via the status report and other vehicles in the project's Communications Plan, this helps ensure that all project stakeholders clearly understand the current state of the project.

At the end of the project, Scope Verification supports the transition of the project to ongoing operations as well as closure of any open contracts or subcontracts. Here again the WBS is used as the basis for the verification and as a key input to the contract and project closure processes.

CHAPTER SUMMARY

This chapter details the transition of the Work Breakdown Structure from its use as a planning tool to its active role as the basis for project execution, management and control.

The initial section of this chapter discusses how the WBS aids in project team acquisition—staffing the project. The next several sections detail how the WBS provides a baseline for project scope management and supports the execution of other project management control processes. The section on the project management triple constraint provides valuable insight into the interplay between time, cost and scope. The chapter concludes with a discussion of the WBS as a basis for Scope Verification.

REFERENCE

Project Management Institute. (2004). *A Guide to the Project Management Body of Knowledge (PMBOK® Guide)—Third Edition.* Newtown Square, PA: Project Management Institute.

CHAPTER QUESTIONS

1. How often are the WBS and WBS Dictionary utilized to verify and validate that the appropriate resources are available and assigned to the project on what basis?
   a. Once
   b. Continually
   c. Never
   d. Sometimes
2. The WBS Dictionary is used to provide a detailed explanation of each ________________.
   a. Activity
   b. Task
   c. Deliverable
   d. Milestone
3. Place the following steps in order

____	Update WBS Dictionary
____	Update Project Budget
____	Update Project Scope Statement
____	Update all planning documents
____	Update Work Breakdown Structure

4. Change Requests should be evaluated against which baseline in order to determine impact to scope?
   a. Project Management Plan
   b. Project Charter
   c. Project Schedule
   d. Work Breakdown Structure and Dictionary
5. Which of the following are not included in the WBS Dictionary?
   a. Acceptance criteria
   b. Completion criteria
   c. Quality measures
   d. Test cases

Chapter 9

Ensuring Success through the WBS

CHAPTER OVERVIEW

As project leaders, one of the questions continually asked of us is, "How is the project doing?" In addition, we are asked to make decisions regarding our project with only as much information as we have available to us. This chapter will cover these topics:

- Project Performance Management, including Scope, Schedule, Cost and Planned versus Actuals
- Stakeholder Management

In order to make sound decisions regarding our projects, we as Project Managers must know what is happening within the project at any given point in time—and we must be aware of that information on a number of different levels. While we may field questions about detailed activities such as, "Has Pat completed the planned 23 lines of code today?" or "When did Sandy finish the Project Charter?" We also receive questions about the larger project perspective. Frequently, Project Managers must answer questions such as, "Where are we against the planned schedule—ahead or behind?" "Where are we against our baselined budget—how much do we have left to spend?" "How are we doing at managing our issues and risks—are we letting them hit us or are we heading them off at the pass?"

PROJECT PERFORMANCE MANAGEMENT

These latter questions can be summarized as questions about project performance management and reporting. Performance reporting is defined as "the process of collecting and distributing performance information. This includes status reports, progress measurement and forecasting". In addition, performance reports are defined as "documents and presentations that provide organized and summarized work performance information, earned value management parameters and calculations, and analyses of project work progress and status" (*PMBOK® Guide*—Third Edition, p. 366).

So, again you may ask, how does the WBS fit into the project in this context? And in truth, the answer is . . . very easily and smoothly! The WBS should be the basis, along with the Scope Statement, the Project Schedule and the Cost Management Plan, for the Performance Measurement Baseline—and should reveal the answers to those difficult questions. As the WBS and WBS Dictionary form the basis from which the Scope Statement, the Project Schedule and the Cost Management Plan are built, they provide the source for metrics used to measure project performance. The Performance Measurement Baseline is defined (*PMBOK® Guide*—Third Edition, p. 366) as "an approved integrated scope-schedule-cost plan for the project work against which project execution is compared to measure and manage performance. Technical and quality parameters may also be included."

Once the scope, schedule, and budget for the project have been baselined, the project team can measure performance against it. Project teams occasionally make this more difficult than need be by making the results something they are not. For example, teams may endlessly analyze data without arriving at conclusions or may over-engineer process measurement. One caution: As the project evolves through approved change requests or the acceptance or mitigation of issues and risks, the control parameters within the performance measurement system must be modified and updated as well. Again, this can be driven from the changes made to the WBS and WBS Dictionary as part of these processes.

So, in detail, how should a project manager use the WBS to help manage performance within a project? Let's break each piece down a little further.

SCOPE

The WBS can be thought of as the backbone of the project. It is initially used as a tool to determine and document scope (see Chapter 2). In addition, it is used to detail the Scope Statement, contracts and agreements (see Chapter 3). But, the WBS has an even bigger impact on the project when it comes to Change Management (see Chapter 8). In managing the performance of the project, the WBS can be used as a simple tool for documenting which deliverables have been completed and which are yet to complete. This provides a summary for the Project Manager of the progress of the project—of how much has been accomplished and how much remains to be accomplished. A drawback to this is that the WBS can be perceived as one-dimensional—all deliverables appear to be the same size and same level of importance. The magnitude of what is completed and what is yet to be finished is difficult to determine.

SCHEDULE

As described in Chapter 7, the WBS is a predecessor or input to the Project Schedule. The deliverables in the WBS are decomposed into Work Packages which are themselves decomposed into tasks, activities and milestones to be completed within the project, defined by the Project Schedule. When the WBS and hence the scope is updated during the project life cycle, the schedule is updated accordingly. For Performance Reporting, the schedule adds a significant component and reference point that provides a dimension of clarity that the WBS alone cannot provide. The schedule allows for the definition of the size and complexity of the effort contained in the Work Packages and defined further in the tasks, activities and milestones present in the schedule. This decomposition and resulting schedule is used to clarify what is and what is not completed at a very detailed level in order to provide a clear picture of the project. It can also provide, again assuming the schedule is accurately maintained and updated on a timely basis, what is remaining to be done, how long it will take to complete (forecasted completion) and a picture of the resources yet to be consumed in the completion of the project. Finally, if the Project Schedule is baselined, it can provide information regarding performance (both schedule and cost) against baseline. This will allow the project manager to determine if the project is ahead or behind schedule and budget (using Earned Value Management (EVM)).

Earned Value Management provides one of the most effective methods for evaluating, then Monitoring and Controlling project progress. By carefully and methodically analyzing planned progress and planned project spend against actual progress and actual project spend, the project manager produces a very clear picture of a project's planned, current and anticipated value. He or she can determine if the progress (productivity) being delivered by the team is delivering value (producing outcomes) at the predicted rate, or at a rate that is faster or slower than planned. The project manager is also able to determine if the rate of performance is being achieved at the anticipated cost, or at a higher or lower cost than was originally planned. Most importantly, using Earned Value Management enables the project manager to predict the eventual cost and delivery date of the project. Although most project management Monitoring and Controlling techniques provide information about a given project's present state or recent past, EVM goes beyond that, providing insight into what will occur in the future. Like a headlight on a train, EVM illuminates the track ahead. Most other project management tools perform like a car's rear-view mirror.

COST

Last, but certainly not least: As it is defined, the WBS provides the basis for estimating and defining the project budget at the work package level. As costs are incurred, actuals will be recorded against the baseline budget. Each Work Package in the WBS is the lowest level of decomposition for the work. The individual Work Packages are also the point at which costs for the work are controlled. As Control Accounts, these elements are the collection point for all costs associated with the work performed.

With this in mind, the WBS Work Packages perform an additional critical role. They provide a natural collection point for the Project Manager to gather costs related to the work being performed, segmented by the structure of the WBS and managed as complete deliverables. Because the WBS can be employed in this way, the Project Manager has a ready breakdown of the cost structure for the project that can be managed along with the work at any level he or she chooses—including the individual Work Package level. As work is performed, the costs associated with the tasks, activities and milestones can be assigned and rolled up to be included in the corresponding individual Work Packages. These in turn can be aggregated and summarized to provide a complete analysis of the entire project cost that is synchronized with the Work Breakdown Structure hierarchy. Costs associated with work that is not-contracted or considered "in house" will be rolled up to "in house" WBS elements (Work Packages). At the same time, costs associated with contracted work will be rolled up to WBS elements representing contracted work in the WBS (Work Packages). In either case, the Project Manager has the ability to manage the costs associated with work defined in the WBS as complete deliverables—either provided entirely by the internal project team, or contracted to another delivery organization.

Other items that are typically included in performance reporting and management are quality control measures and technical parameters. From a quality perspective, control should be defined at each work package level from the WBS. Actual results can then be measured and reported against the defined baseline. The same is true with technical parameters. Once these items are baselined, they do not change, except in response to approved change requests. After a change has been reviewed and approved, all affected plans must be re-baselined to ensure that measures and metrics (actuals) are being collected and compared to the correct baseline.

PLANNED VERSUS ACTUAL

Now what about comparing actual performance to these baselines? What is the source for the actual data? These data can come from a variety of sources—project status meetings, status reports, time systems, budgeting and cost systems, and less reliably, word of mouth.

During the Execution of the project and as part of Monitoring and Controlling, it is essential to report results regularly, accurately and with the appropriate detail to meet the needs of the stakeholders who will review it. To do this you will perform an analysis that compares the planned version of your project (baselines) with the actual results you have available at any given point in time. It may be trend, variance, earned value—each of these is a comparison between what is currently being delivered and the original baseline. Such detailed analysis is the subject for another book—but, no matter what the circumstance, you *must* report results!

To do this, you must organize the information and make it presentable to the audience. Presenting raw data is not an effective communication methodology for your stakeholders, so you will need to sort, sift, format and present the information in a logical manner. The information must be readable, understandable and meaningful to the parties receiving it. What better way to organize the data than in a way that has been previously shared—in the order that the WBS is organized? It does not have to be a one-to-one match, but the WBS is organized in a fashion that reveals how the work is grouped—why not organize the information to be presented in a similar manner?

Once the data are defined, gathered, analyzed, summarized and documented, they should be presented to stakeholders along with recommendations from the Project Manager regarding actions that may be taken based on existing variances from baseline or potential positive variances.

Based on the information presented during project reviews, executives, business sponsors and leaders review the materials and recommended courses of action and, together with the Project Manager, determine a forward direction. If it is determined that a change or corrective action must be taken, those changes will be sent through the applicable project control processes while ensuring all appropriate project assets and documentation (including the WBS and WBS Dictionary) are updated.

STAKEHOLDER MANAGEMENT

In the same manner that performance management and reporting are used to inform project sponsors and executives, there are many other project stakeholders who need key information about the project. These individual stakeholders and stakeholder groups can be defined as those who are "impacting or are impacted by a project" (Stakeholder Definition, Advanced Strategies, Inc., http://www.advstr.com/web/default.cfm). Although these stakeholders can sometimes be overlooked, it is important for the Project Manager to not underestimate the voice and influence of this group. This is an area where a Communication Plan aligned with and traceable to the WBS can be invaluable. By carefully identifying the stakeholders during the development of the Communication Plan, the Project Manager can ensure that all stakeholders are informed and receive "actionable" information from which they can make key business decisions.

CHAPTER SUMMARY

In summary, project performance and stakeholder management are critical areas of interest in the project life cycle. By utilizing defined project performance management metrics, a Project Manager can determine if the project will be on-time, within budget and able to deliver to the customer the deliverables they need and expect. Coincident with the utilization of performance metrics, performance results should be shared with stakeholders who are impacted by the project's progress.

As the project proceeds through its life cycle, performance metrics provide decision support information stakeholders require to evaluate the contribution of the project to the business. Additionally, at the tactical level, this information offers insight for the Project Manager and team into the readiness to progress into the next phase of the project, Closeout.

REFERENCES

Advanced Strategies, Inc. 1988. Stakeholder Definition, Retrieved July 2007 from http://www.advstr.com/web/default.cfm.

Project Management Institute. 2004. *A Guide to the Project Management Body of Knowledge (PMBOK® Guide)—Third Edition.* Newtown Square, PA: Project Management Institute.

CHAPTER QUESTIONS

1. Which of the following is *not* a source from which metrics used in project performance are derived?
   a. Issues Log
   b. WBS and WBS Dictionary
   c. Project Schedule
   d. Cost Management Plan
2. What is one of the drawbacks of the WBS?
   a. It has to be updated.
   b. It is hard to create.
   c. It can be perceived as one-dimensional.
   d. It can't be done on paper.
3. Before a Project Schedule can be used for performance reporting, what final step must be taken?
   a. Create the schedule.
   b. Update the schedule.
   c. Add the schedule to a project scheduling tool.
   d. Baseline the schedule.
4. Which project management analysis technique enables the project manager to predict the cost and delivery date of the project based on its performance over time?
   a. Cost Management
   b. Work Breakdown Structures
   c. Earned Value Management
   d. Project Scoping
5. Which is a recommended method for organizing project data for presentation to stakeholders?
   a. Grouped alphabetically
   b. Grouped chronologically (by schedule)
   c. Grouped by Cost Account
   d. Grouped by Work Breakdown Structure hierarchy

Chapter 10

Verifying Project Closeout with the WBS

CHAPTER OVERVIEW

This chapter will cover the Closing phase of the project life cycle and the use of the Work Breakdown Structure in this important phase. Project Closeout includes ensuring that all deliverables in scope have been completed prior to close, completing acceptance and turnover of the product, preparing for support and maintenance of the project once completed and contract closure. Included in this chapter on project closure are discussions on the following topics:

- Acceptance/Turnover/Support/Maintenance
- Contract Closure
- Project Closeout

PROJECT CLOSEOUT

During Project Closeout it is imperative that the Project Manager verify that all of the deliverables produced by contracted organizations are delivered in accordance with contract terms. Again, as changes have been made to the project, the contracts with each group should have been maintained in lock step with updates to the WBS and the WBS Dictionary. By reviewing each contract and the deliverables associated with them, the Project Manager can determine if all work has been completed by the particular contracting organization. Verifying completion by contracting organizations becomes extremely important as work is delivered and contractors are looking for contract finalization and payment—more on this later. Again, if the deliverables do not align with the approved criteria in the WBS and WBS Dictionary, there is still more work to be done. If, by contrast, the deliverables do conform to quality and acceptance criteria, documentation regarding the completion—including sign-off for all contracted deliverables will be completed by the Project Manager and contractor—and should be placed in the contracts file for future reference. The project has then reached an important milestone—all internal closeout activities are complete!

If the project team has completed all its work up to this point this should be the easiest part—final sign off of the product delivery by the customer. At this point, the Project Manager and project team should be able to review and explain the original scope of the project, including the deliverables to be created. This can be traced through the Project Charter, Scope Statement, contracts and agreements. As changes were made during the project (via change management), all of these items, including the WBS, were updated. By comparing the planned deliverables against what was actually produced and delivered by the project team, the review, approval and sign-off with the customer should be easy and straightforward. The Project Manager can easily demonstrate and compare what was defined by the stakeholders—to what was actually delivered.

ACCEPTANCE / TURNOVER / SUPPORT / MAINTENANCE

So, what if the customer organizations or teams don't like what they are receiving? What if they do not agree that what is being provided is what they asked for? What if they refuse to make final payment on the delivered product? The agreed-upon Project Charter, Scope Statement, WBS, WBS Dictionary, contracts and agreements are the starting points for negotiation about the acceptability of the delivered product.

As part of these discussions, the future production support and warranty period (if applicable) for the delivered product should once again be reviewed for appropriateness and final sign-off.

CONTRACT CLOSURE

Following Scope Verification and agreement between the Project Manager, project team and the receiving organization, the next step is to close all outstanding contracts. At this point, the Project Manager can review each contract for remaining work in order to close the associated contract(s). Documentation will be placed in the project contract file that the work has been completed satisfactorily and final payments for the work can be approved. So, now the project is complete . . . correct? Actually, now the project "close out" begins.

PROJECT CLOSEOUT

Once all contracts have been closed and all customer sign-offs have been completed, the Project Manager still has quite a bit of work to do.

First, the Project Manager initiates a post-review of the project with an effort to capture and document lessons learned. You may be thinking—what is this doing in a book about Work Breakdown Structures? Here, the WBS is used as the basis for project post review—as it organizes the scope of work in a logical manner. By again using the WBS to guide project post review, the project team can focus on the review of the work as it is defined by the WBS. The segmentation of the work as it appears in the WBS enables review of individual deliverables systematically as opposed to reviewing the outcomes of the entire project at once. Post-review should be conducted for each of the individual project deliverables and should be documented and stored for future reference.

During the project post-review, all documentation for the project should be updated to record and reflect final results. Once all of these activities have been completed, all documentation from the WBS through the project post review is stored in a document repository accessible by the sponsor, project team and stakeholders. Performing this critical final step allows for the retrieval of all historic information regarding the project as well as sign-offs and final documented delivery. Project teams are then encouraged to retrieve and reference this information, extracting key documents that can be used as templates for future efforts. As an example, the WBS may be referenced in the future as a basis for a revised tax assessment (if needed) or for capitalizing work that has been completed.

CHAPTER SUMMARY

As is evidenced in all of the chapters in this book, the WBS and WBS Dictionary are vitally important during *all* phases of the project lifecycle and are a critical foundation and guiding tool for ensuring successful project delivery.

CHAPTER QUESTIONS

1. The WBS and WBS Dictionary are vitally important during *all* phases of the project lifestyle.
   a. False
   b. True
2. Which of the following are *not* utilized during negotiations about acceptability of the delivered product(s)?
   a. WBS
   b. Contracts / Agreements
   c. WBS Dictionary
   d. Risk Register
3. Changes to the baselined Work Breakdown Structure should be made through which project process?
   a. Project Planning
   b. Change Management
   c. Scoping
   d. Project Closeout
4. What is the first step in project Closing activities?
   a. Celebrate.
   b. Contract closure.
   c. Initiate a project post-review.
   d. Verify all deliverables have been completed.
   e. Update all documentation to record and reflect final results.
5. The work breakdown structure can be utilized as basis for project post-review.
   a. True
   b. False

Part III

WBS For Project Management Decomposition

Chapter 11

A Project Management WBS

CHAPTER OVERVIEW

For many years we, your authors, have focused on expanding the awareness and application of Work Breakdown Structures while sharing current accepted practice for developing them. Over this time, each of us has received comments, feedback, suggestions and repeated questions from project management practitioners, asking if we knew of existing examples that could be used as templates for detailing the breakdown of the Project Management components of the WBS. In answer to those who have inquired, and for those who have quietly thought about this subject but never asked, we have included Chapter 11. This chapter is devoted entirely to providing examples of the Project Management components of any WBS.

The major sections of this chapter include the following:

- Organization Options for a Project Management WBS
- Project Management WBS Aligned with the *PMBOK® Guide*—Third Edition
- Project Management WBS "lite"

Before going further, however, we would like to address a few concerns you may have. First and foremost, there is no *single* right way of representing the Project Management components of a WBS. As we've stated numerous times throughout this book, the right representation for your project depends entirely upon your needs. Yes, this is yet another

Use-Related Characteristic. Next, the middle section of this chapter provides an example of Project Management WBS components aligned with our review of PMI's *PMBOK® Guide*—Third Edition. While we do believe we have a reasonable level of knowledge about Work Breakdown Structures, we do not profess to possess expert knowledge about the *PMBOK® Guide*—Third Edition. Considering this, the Project Management component WBS representations presented in this section are based solely on our review and interpretation of this global standard, and should not be interpreted as sanctioned PMI work. Finally, the third and last section of this book presents a "lite" representation of the Project Management components of the WBS based on our applied experience and reflects what we believe applies to most projects, most of the time.

ORGANIZATION OPTIONS FOR A PROJECT MANAGEMENT WBS

Project Management as a component of work reflected in any project is a difficult area to decompose as part of a Work Breakdown Structure because it implies both discrete and level-of-effort deliverables. Examples of discrete deliverables include a Scope Statement, the WBS itself, a Project Schedule, and many other project management deliverables too numerous to note. By its very nature, project management can *also* be considered a level-of-effort deliverable, characterized by a uniform rate of work (project management practice) over a period of time (the project's duration). This difficulty is compounded further given the two different ways in which project management is broken down in the *PMBOK® Guide*—Third Edition.

When you read this global standard, you quickly see that the practice of project management is broken down both by *Process Group* and *Knowledge Area*. The *PMBOK® Guide*—Third Edition breaks down project management processes and deliverables into five different *Process Groups*:

- Initiating
- Planning
- Executing
- Monitoring and Controlling
- Closing

This common and familiar grouping of project delivery is represented in the *PMBOK® Guide*—Third Edition to reflect how projects are actually performed. First there is Initiating, then Planning followed by Executing, Monitoring and Controlling and Closing. It follows then, that to represent this Project Management work appropriately beginning with the Level 2 Project Management element of the WBS, the decomposition of the Project Management deliverables would follow this organization. The **Process Group** breakdown of the Project Management deliverables is depicted in Exhibit 11.1:

1 Project Management
- 1.1 Initiating
- 1.2 Planning
- 1.3 Executing
- 1.4 Monitoring and controlling
- 1.5 Closing

Exhibit 11.1 Project Management WBS by Process Group

Figure 11.1 shows this same Project Management component of the WBS with the *Process Group* orientation, only depicted in a hierarchical, inverted tree structure view placed within the context of the full project.

Also represented in the *PMBOK® Guide*—Third Edition are the **Knowledge Areas** that make up the Body of Knowledge, the core information about the practice of Project Management. Each of the *Knowledge Areas* are described in separate chapters, one devoted to

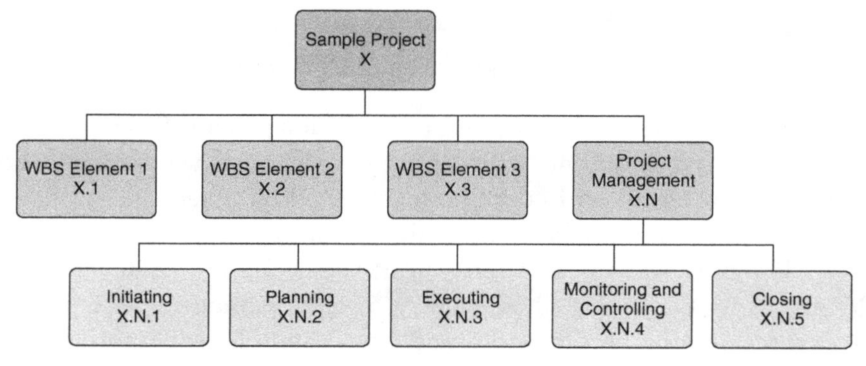

Figure 11.1 Project management WBS by process group top level view.

each *Knowledge Area* that follow the "Project Management Standard" described in Chapter 3. These Knowledge Areas are as follows:

- Integration Management
- Scope Management
- Time Management
- Cost Management
- Quality Management
- Human Resource Management
- Communications Management
- Risk Management
- Procurement Management

If our Project Management components of the WBS level 2 element were to be decomposed in this manner, the resulting levels of the WBS would be as represented in Exhibit 11.2:

1 Project Management
- 1.1 Integration Management
- 1.2 Scope Management
- 1.3 Time Management
- 1.4 Cost Management
- 1.5 Quality Management
- 1.6 Human Resources Management
- 1.7 Communications Management
- 1.8 Risk Management
- 1.9 Procurement Management

Exhibit 11.2 Project Management WBS by Knowledge Area

Figure 11.2 shows this same Project Management WBS with the knowledge area orientation, only depicted in a hierarchical, inverted tree structure view placed within the context of the full project.

While both representations are valid, we believe that the *Knowledge Area* view may work best when you consider the day-to-day job of a Project Manager. As noted previously, project management is unique in that it contains both discrete and level-of-effort deliverables. Many project management deliverables are used (created, updated, reviewed,

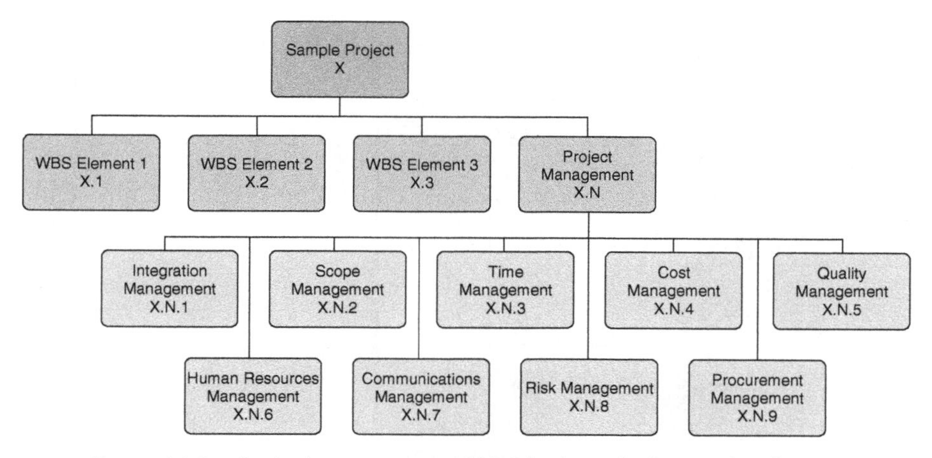

Figure 11.2 Project management WBS by knowledge top level view.

etc.) throughout the project management lifecycle. As such, they can logically be placed in multiple process groups. Given this, we find it most appropriate and simpler to represent the Project Management deliverables only once.

This would be more easily done if the decomposition is by Knowledge Area. Reflecting on the Core Characteristics we discussed earlier, and given that a deliverable must be represented only once in any WBS, decomposing the WBS by *Knowledge Area* would help support the quality of the WBS as well.

The next two illustrations detail two possible representations of the Project Management components of a Work Breakdown Structure. The first illustration presents an example that is closely aligned with the *PMBOK® Guide*—Third Edition. The second is a "lite" version of the same information. Either of these representations will apply to most projects, most of the time.

PROJECT MANAGEMENT WBS COMPONENTS ALIGNED WITH THE PMBOK® GUIDE—THIRD EDITION

Table 11.1 depicts an example of the Project Management components of the WBS closely aligned with our review of PMI's *PMBOK® Guide*—Third Edition. This decomposition is organized both by project management *Process Group* and by *Knowledge Area*. In this view we also ensure that

Table 11.1 *PMBOK® Guide*—Third Edition Aligned Project Management WBS

WBS Level	WBS Code	WBS Element Title	Description (*PMBOK® Guide* Glossary or Third Edition Applicable Errata)	Process Group				
1	x	**Project Name**	**Entire project scope including all other project deliverables. Represents 100% of the project scope.**	**Initiating**	**Planning**	**Executing**	**Monitoring & Controlling**	**Closing**
2	*x.n*	**Project Management**						
3	x.n.1	**Project Integration Management**	**The processes and activities needed to identify, define, combine, unify, and coordinate the various processes and project management activities within the Project Management Process Groups.**					

4	*x.n.1.1*	Contract	A contract is a mutually binding agreement that obligates the seller to provide the specified product or service or result and obligates the buyer to pay for it.	x		x	x	x
4	*x.n.1.2*	Project Statement of Work		x				

Source: Project Management Institute, A Guide to the Project Management Body of Knowledge (*PMBOK® Guide*—Third Edition). PMI. Newtown Square: PA.

the *Knowledge Area* view additionally shows the *Process Group(s)* to which each of the individual deliverables belong. We have included this alignment to help clarify how each of the deliverables relates both to the Knowledge Areas of the *PMBOK® Guide*—Third Edition, as well as the *Process Groups*. We also want to explain what you will find in each of the columns in this illustration.

This WBS decomposition is represented in a Microsoft Excel Workbook format. We have chosen this type of illustration for a number of reasons. Most importantly, this representation facilitates the use of the same information in other formats. What we mean is that the information you find in this version of the WBS can be easily extracted or converted to representations that would be appropriate for communicating with other project stakeholders. For instance, you may wish to represent the WBS as an outline or an organization chart graphic. To do this, the "workbook" information can be easily copied and pasted into either of those formats.

This format also allows the inclusion of other key information, particularly, the information found in the WBS Dictionary. Columns can be added to the workbook format to append detail to each of the WBS elements. This also allows the direct correlation between the WBS element and its WBS Dictionary explanation.

For your use, we have provided a WBS template, containing the decomposition orientations we've described (see Appendix 3 for the entire documents). The first breakdown represents the Project Management deliverables in the *Process Group* orientation. The second breakdown contains the *Knowledge Area* decomposition. These two decompositions of the Project Management deliverables contain precisely the same information. They only differ in their orientation . . . one by *Process Group*, the other by *Knowledge Area*. We use the Knowledge Area representation for our illustration in Table 11.1. The columns you will find in the workbook and illustration are (from left to right): WBS Breakdown Level, WBS Code, WBS Element Title, Description (in this case we've included WBS Dictionary information from the *PMBOK® Guide*—Third Edition) and *Process Group* alignment.

PROJECT MANAGEMENT WBS LITE

Table 11.2 depicts the "lite" representation of the same Project Management components in a WBS. If the detail provided by the fully elaborated

Table 11.2 Project Management WBS "Lite"

WBS Level	WBS Code	WBS Element Title	Description (*PMBOK*® *Guide*—Third Edition Glossary or Applicable Errata)
1	x	**Project Name**	**Entire project scope including all other project deliverables. Represents 100% of the project scope.**
2	*x.n*	**Project Management**	
3	x.n.1	**Project Integration Management**	**The processes and activities needed to identify, define, combine, unify, and coordinate the various processes and project management activities within the Project Management Process Groups.**
4	*x.n.1.1*	Contract	A contract is a mutually binding agreement that obligates the seller to provide the specified product or service or result and obligates the buyer to pay for it.
4	*x.n.1.2*	Project Statement of Work	

Source: Project Management Institute, A Guide to the Project Management Body of Knowledge (*PMBOK*® *Guide*—Third Edition). PMI. Newtown Square: PA.

example and template is overkill, this "lite" version may suit your needs more adequately.

CHAPTER SUMMARY

This chapter provides some guidance for a topic we have been wrestling with for quite a long time. Over the past years, we have received many requests and questions about the most effective way to represent Project Management work elements (deliverables) within the Work Breakdown Structure, while ensuring it is done in a manner that is meaningful to a majority of the project's stakeholders. This chapter addresses that challenge.

The first section of the chapter provides some information to ground the discussion of the Project Management deliverables we include. Following that discussion we review two key approaches to the Project Management WBS representation; decomposition by *Process Group* and decomposition *Knowledge Area*. We have chosen to discuss both approaches to reflect the organization of Project Management practice as it is discussed in the *PMBOK® Guide*—Third Edition.

The final section of the chapter illustrates the Project Management components of a WBS directly aligned with the tools, techniques and topics found in the *PMBOK® Guide*—Third Edition. Some project managers find this level of detail and decomposition to be overkill, or too much information considering the relatively small size of their projects. Their claim is that the Project Management deliverables component of their WBS would then be more extensive than the remainder of the entire project. Because of this, we have also included an illustration of the Project Management components represented in what we would call Project Management component deliverables "lite."

Please remember these illustrations represent your authors' views about how to present a fully decomposed as well as a "lite" version. These particular approaches may or may not be appropriate for your specific project(s).

A FINAL WORD

Our hope is that you find the information included in the chapters, quizzes and appendices in this book to be as helpful to you as it was educational and enjoyable for us to develop. The last year has been very enlightening for the three of us and we hope we have shared tips and techniques that you will find useful – now or in the future.

REFERENCE

Project Management Institute. 2004. *A Guide to the Project Management Body of Knowledge (PMBOK® Guide—Third Edition)*. Newtown Square, PA: Project Management Institute.

Appendix A

Project Charter Example

PROJECT OVERVIEW

This project is being undertaken to establish a new residence for Mr. and Mrs. John Smith. The new residence will be a free-standing, single-family dwelling built on a two-acre lot (lot #24) located at 200 North Maple Avenue, MyTown, MyState, 20001-1234, USA. The project is to commence on Monday, February 2, 2015 and will complete no later than Thursday, December 31, 2015.

This home is being constructed to take advantage of the latest building materials and codes and will employ emerging technology to minimize energy consumption. Construction will be overseen and managed by Apex Home Builders, the prime contractor who may subcontract components of the construction effort.

All labor will be bonded and all materials will meet or exceed local building code guidelines.

SECTION I. PROJECT PURPOSE

The home project is being undertaken to establish a new primary residence for Mr. and Mrs. Smith and family. The new residence is scheduled for completion in December so that the Smith family may move in during the first two weeks of 2016. Mr. Smith will be taking responsibility for his company's North American operations in 2016 and is relocating from Europe to do so. Mr. Smith and family will be traveling and relocating during December of 2015 and will move directly from their current home to the newly completed residence.

The home must be completed by December 31, 2015 so that the Smith's can establish residence in the community with the appropriate lead time to enable their children to be enrolled in the school system to begin the 2016 school year along with their classmates.

SECTION II. PROJECT SCOPE

This is a Fixed-Price Contract

Contractor commitment estimate is U.S. $750,000.00.

Upon completion, the new property will include the following as described in the detailed specifications and blueprint:

- Landscaping
- Foundation (with basement)—poured concrete and concrete block
- Driveway—2000 feet, concrete with brick inlay
- Main Home—4500 square feet, brick/stucco
- Deck / Patio / Screen Room
- Garage—1600 square feet, two story

SECTION III: PROJECT OBJECTIVES

As described in Section II, completion of the project must be achieved by December 31, 2015. Progress milestones associated with the project are as follows:

1. Architectural drawings complete and approved
2. Building permit approved
3. Lot preparation and clearing complete
4. Foundation excavation complete
5. Footings poured and set
6. Foundation poured, block construction complete, foundation set
7. Home and garage exterior closed to weather
8. Driveway and landscape complete
9. Interior wiring complete
10. Exterior wiring complete
11. HVAC complete
12. Interior plumbing complete
13. Exterior plumbing complete
14. Interior finish complete

15. Exterior finish complete
16. Walkthrough complete
17. Certificate of Occupancy granted
18. Interior and exterior punch list approved
19. Interior and exterior punch list complete
20. Acceptance review and key turnover complete

SECTION IV: OUTSTANDING ISSUES

- Prime contractor will be responsible for all work and workmanship
- Incentive approved for early delivery—5%
- Penalty approved for late delivery—5%
- Temperature expected to average below 20°F beginning September 15, 2015

SECTION V: APPROVALS

All funding has been preapproved and placed in a reserve account against which the primary contract or may draw. Contractor may draw quarterly payments of equal amount beginning with the second quarter following start of work. Buyer will withhold first quarterly payment until all work is completed and approved at end of project close and turnover. Pending satisfactory acceptance of project close and transition, buyer will provide final payment to contractor.

Note: Invoices will be generated by contractor at the close of each construction quarter.

- Invoicing will document all materials purchased during the previous quarter.
- Invoicing will document all work completed during the previous quarter.
- Payment will be made in four (4) equal quarterly payments.
- Anticipated materials or labor cost variance above 2% in any single quarter, or 5% overall requires detailed explanation and approval of the buyer before commitment to purchase or perform work.
- All cost increases invoiced without prior approval either as defined by the previous bullet or by the buyer will be paid at the original contract price.

SECTION VI: REFERENCES

- See applicable state and local building codes.
- Site plan and building permit files at City Court, Mytown, Mystate.

SECTION VII: TERMINOLOGY

N/A

SECTION VIII: PROJECT APPROACH

- Prime contractor will maintain all project documents and schedule.
- Prime contractor will perform work with contractor's own employees who are bonded and hold the appropriate trade licenses and credentials. In the event that the prime contractor subcontracts work, all subcontractor's employees will be verified by prime contractor to be bonded and will hold the appropriate trade licenses and credentials.
- Prime contractor agrees to pay a 10% single quarter penalty for any project worker found on the project site without proof of proper trade license and credentials.
- Prime contractor will be responsible for all subcontract oversight, deliverables and management.
- Prime contractor has agreed to the terms of the fixed-price contract and schedule.
- During planned project execution, prime contractor will provide periodic progress reports to buyer (biweekly, monthly).
- Following any schedule delay, prime contractor agrees to provide progress reports on a weekly basis until schedule has been caught up.

SECTION IX: PROJECT DELIVERABLES AND QUALITY OBJECTIVES

- See Section II and Section III.
- All HVAC (heating, ventilation and air conditioning), landscaping, finish work, including inside and exterior doors, interior and exterior cabinetry and doors and door and cabinetry hardware, bathroom fixtures, kitchen appliances, counter tops, backsplash material, garage finish work, flooring and attic are at the contractors "Level V—Premium Grade" per the contractor's detailed house plans and upgrade options.

- Buyer will select bathroom fixtures, kitchen and laundry room fixtures and appliances as specified on contractor's detail schedule.
- Contractor has included a $10,000 lighting allowance. All lighting requirements that exceed the allowance will require submittal of change request to the contractor.

SECTION X: ORGANIZATION AND RESPONSIBILITIES

- See Section VIII.

SECTION XI: PROCESS OPTIONS AND DEVIATIONS

- See Section IV.

SECTION XII: QUALITY CONTROL ACTIVITIES

- See Section IV.
- Contractor reserves the right to require change orders for any/all modifications to current design and "Level V—Premium Grade" finish plans following contract approval. Additionally, contractor reserves the right to require change orders for any/all modifications to project (plan, design, schedule, scope) requested by purchaser following official schedule approval prior to commencement of work.

SECTION XIII: PROJECT SCHEDULE

- See Section I, Section VIII.
- Prime contractor will provide detailed schedule prior to the start of project.

SECTION XIV: PROJECT EFFORT ESTIMATE

- See Section I.

SECTION XV: PROJECT COST ESTIMATE

- See Section II.

Appendix B

Project Scope Statement Example

PROJECT OVERVIEW

This project is being undertaken to establish a new residence for Mr. and Mrs. John Smith. The new residence will be a free-standing, single-family dwelling built on a two-acre lot (lot 24) located at 200 North Maple Avenue, MyTown, MyState, 20001-1234, USA. The project is to commence on Monday, February 2, 2015 and will complete no later than Thursday, December 31, 2015.

This home is being constructed to take advantage of the latest building materials and codes and will employ emerging technology to minimize energy consumption. Construction will be overseen and managed by Apex Home Builders, the prime contractor who may subcontract components of the construction effort. All labor will be bonded and all materials will meet or exceed local building code guidelines.

SECTION I. PROJECT PURPOSE

The home project is being undertaken to establish a new primary residence for Mr. and Mrs. Smith and family. The new residence is scheduled for completion in December so that the Smith family may move in during the first two weeks of 2016. Mr. Smith will be taking responsibility for his company's North American operations in 2016 and is relocating from Europe to do so. Mr. Smith and family will be traveling and relocating during December of 2015 and will move directly from their current home to the newly completed residence.

The home must be completed by December 31, 2015, so that the Smith's can establish residence in the community with the appropriate lead time to enable their children to be enrolled in the school system to begin the 2016 school year along with their class mates.

SECTION II. PROJECT SCOPE

This is a Fixed-Price Contract

Contractor commitment estimate is U.S. $750,000.00.

Upon completion, the new property will include the following as described in the detailed specifications and blueprint:

- Landscaping
- Foundation (with basement)—poured concrete and concrete block
- Driveway—2000 feet, concrete with brick inlay
- Main home—4500 square feet, brick/stucco
- Deck / Patio / Screen Room
- Garage—1600 square feet, two story

SECTION III: PROJECT MILESTONES

As described in Section II, completion of the project must be achieved by December 31, 2015. Progress milestones associated with the project are as follows:

1. Architectural drawings complete and approved
2. Building permit approved
3. Lot preparation and clearing complete
4. Foundation excavation complete
5. Footings poured and set
6. Foundation poured, block construction complete, foundation set
7. Home and garage exterior closed to weather
8. Driveway and landscape complete
9. Interior wiring complete
10. Exterior wiring complete
11. HVAC complete
12. Interior plumbing complete
13. Exterior plumbing complete
14. Interior finish complete
15. Exterior finish complete

16. Walkthrough complete
17. Certificate of Occupancy granted
18. Interior and exterior punch list approved
19. Interior and exterior punch list complete
20. Acceptance review and key turnover complete

SECTION IV: PROJECT APPROACH

- Prime contractor will maintain all project documents and schedule.
- Prime contractor will perform work with contractor's own employees who are bonded and hold the appropriate trade licenses and credentials. In the event that the prime contractor subcontracts work, all subcontractor's employees will be verified by prime contractor to be bonded and will hold the appropriate trade licenses and credentials.
- Prime contractor agrees to pay a 10% single quarter penalty for any project worker found on the project site without proof of proper trade license and credentials.
- Prime contractor will be responsible for all subcontract oversight, deliverables and management.
- Prime contractor has agreed to the terms of the fixed-price contract and schedule.
- During planned project execution, prime contractor will provide periodic progress reports to buyer (biweekly, monthly).
- Following any schedule delay, prime contractor agrees to provide progress reports on a weekly basis until schedule has been "caught up."

ISSUE MANAGEMENT

- Project-related issues will be tracked, prioritized, assigned, resolved, and communicated in accordance with the Prime Contractor's Issue Management protocol.
- Issues will be reported using an Issue Report Form. Issue descriptions, owners, resolution and status will be maintained in an Issues Log in a standard format.
- Issues will be addressed with the project owner and communicated in the project weekly status report.

CHANGE MANAGEMENT

The change control procedures as documented in the prime contractor's Change Management Plan will be consistent with standard home construction methodology and consist of the following processes:

- The Project Manager will establish a Change Log to track all changes associated with the project effort.
- All Change Orders must be submitted via a Change Order Form and will be assessed to determine possible alternatives and costs.
- Change Orders will be reviewed and approved by the project owner and accepted/acknowledged by the buyer.
- The effects of approved Change Order on the scope and schedule of the project will be reflected in updates to the Project Plan.
- The Change Log will be updated to reflect current status of Change Orders.

COMMUNICATIONS MANAGEMENT

The following strategies have been established to promote effective communication within and about this project. Specific communication policies will be documented in the prime contractor's Communication Plan.

- The Prime Contractor's Project Manager will present project status to the buyers on a biweekly basis.
- The buyer will be notified by the prime contractor via e-mail or telephone of all urgent issues. Issue notification will include time constraints, and impacts, which will identify the urgency of the request.
- The buyer will notify the prime contractor of schedule, scope or budget modifications in a timely manner. Communications of changes may be made by voicemail or telephone, *but will not be acted upon by prime contractor until a Change Order Form is received.*

PROCUREMENT MANAGEMENT

The prime contractor will maintain a Procurement Management Plan in accordance with the Project Plan. The Procurement Plan will document the following:

- How much, when and by what means each of the materials and services that this project requires will be obtained
- The types of subcontracts required (if any)
- How independent estimates (as evaluation criteria) will be obtained
- How procurement will be coordinated with project schedule and budget
- What a subcontracted Statement of Work includes
- Potential sources of goods and services

RESOURCE MANAGEMENT

The Prime Contractor will produce a Resource Management Plan that will document the following:

- All materials and services to be delivered as part of the project along with cost estimates and quality information
- Which materials and services will be obtained from sources outside the Prime Contractor's organization

SECTION V: OUTSTANDING ISSUES

- Prime contractor will be responsible for all work and workmanship
- Incentive approved for early delivery—5% of contract value Penalty approved for late delivery—5% of contract value
- Temperature expected to average below 20°F beginning September 15, 2015

SECTION VI: APPROVALS

All funding has been preapproved and placed in a reserve account against which the primary contractor may draw. Contractor may draw quarterly payments of equal amount beginning with the second quarter following start of work. Buyer will withhold first quarterly payment until all work is completed and approved at end of project close and turnover. Pending satisfactory acceptance of project close and transition, buyer will provide final payment to contractor.

Note: Invoices will be generated by contractor at the close of each construction quarter.

- Invoicing will document all materials purchased during the previous quarter.
- Invoicing will document all work completed during the previous quarter.
- Payment will be made in four (4) equal quarterly payments.
- Anticipated materials or labor cost variance above 2% in any single quarter, or 5% overall requires detailed explanation and approval of the buyer before commitment to purchase or perform work.
- All cost increases invoiced without prior approval either by the previous bullet or by the buyer will be paid at the original contract price.

SECTION VII: REFERENCES

- See applicable state and local building codes
- Site plan and building permit will be filed at City Court, Mytown, Mystate.

SECTION VIII: PROJECT DELIVERABLES AND QUALITY OBJECTIVES

- See Sections I, II and III.
- All HVAC (heating, ventilation and air conditioning), landscaping, finish work, including inside and exterior doors, interior and exterior cabinetry and doors and door and cabinetry hardware, bathroom fixtures, kitchen appliances, counter tops, backsplash material, garage finish work, flooring and attic are at the contractors "Level V—Premium Grade" per the contractor's detailed house plans and upgrade options.
- Buyer will select bathroom fixtures, kitchen and laundry room fixtures and appliances as specified on contractor's detail schedule.
- Contractor has included a $10,000.00 lighting allowance. All lighting requirements that exceed the allowance will require submittal of change request to the contractor.

SECTION IX: QUALITY CONTROL ACTIVITIES

- See Section III, IV.
- Contractor reserves the right to require change orders for any/all modifications to current design and "Level V—Premium Grade"

finish plans following contract approval. Additionally, contractor reserves the right to require change orders for any/all modifications to project (plan, design, schedule, scope) requested by purchaser following official schedule approval prior to commencement of work.

SECTION XI: PROJECT SCHEDULE

- See Section I, Section IV.
- Prime contractor will provide detailed schedule prior to the start of project.

Appendix C

Project Management WBS Examples

This appendix includes decomposition of the project management component of the WBS organized both by project management *process group* and *knowledge area*.

PROCESS GROUP VIEW

Table C.1 depicts the complete example of the Project Management components of the WBS closely aligned with the authors' review of PMI's *PMBOK® Guide*—Third Edition. This view is organized by Process Group.

KNOWLEDGE AREA VIEW

Table C.2 also depicts a complete example of the Project Management components of the WBS closely aligned with the authors' review of PMI's *PMBOK® Guide*—Third Edition. This view is organized by Knowledge Area.

KNOWLEDGE AREA "LITE" VIEW

Table C.3 depicts the complete example of the "lite" representation of the same Project Management components in a WBS. If the detail provided by the fully elaborated example and template is overkill, this "lite" version may suit your needs more adequately.

Table C.1 Process Group View

WBS LEVEL	WBS Code	WBS Element Title	Description (*PMBOK® Guide*—Third Edition Glossary or Applicable Errata)
1	***x***	Project Name	Entire project scope including all other project deliverables. Represents 100% of the project scope.
2	*x.n*	**Project Management**	
3	***x.n.1***	**Initiating**	**This level represents the summary level for the Initiating process group.**
4	*x.n.1.1*	Project Charter	A document issued by the project initiator or sponsor that formally authorizes the existence of a project, and provides the project manager with the authority to apply organizational resources to project activities.
4	*x.n.1.2*	Contract	A contract is a mutually binding agreement that obligates the seller to provide the specified product or service or result and obligates the buyer to pay for it.
4	*x.n.1.3*	Project Statement of Work	
4	*x.n.1.4*	Organizational Process Assets	Any or all process related assets, from any or all of the organizations involved in the project that are or can be used to influence the project's success. The process assets include formal and informal plans, policies, procedures, and guidelines. The process assets also include the organization's knowledge bases such as lessons learned and historical information.
3	***x.n.2***	**Planning**	**This level represents the summary level for the Planning process group.**

4	*x.n.2.1*	Product Scope Description	The documented narrative description of the product scope.
4	*x.n.2.2*	Preliminary Project Scope Statement	
4	*x.n.2.3*	Project Scope Statement	The narrative description of the project scope, including major deliverables, projects objectives, project assumptions, project constraints, and a statement of work, that provides a documented basis for making future project decisions and for confirming or developing a common understanding of project scope among the stakeholders. The definition of the project scope—what needs to be accomplished.
4	*x.n.2.4*	Scope Baseline	
4	*x.n.2.5*	Contract Statement of Work	A narrative description of products, services, or results to be supplied under contract.
4	*x.n.2.6*	Work Breakdown Structure	A deliverable-oriented hierarchical decomposition of the work to be executed by the project team to accomplish the project objectives and create the required deliverables. It organizes and defines the total scope of the project. Each descending level represents an increasingly detailed definition of the project work. The WBS is decomposed into work packages. The deliverable orientation of the hierarchy includes both internal and external deliverables.

(*continues*)

Table C.1 *(continued)*

WBS LEVEL	WBS Code	WBS Element Title	Description (*PMBOK*® *Guide*—Third Edition Glossary or Applicable Errata)
1	*x*	Project Name	Entire project scope including all other project deliverables. Represents 100% of the project scope.
4	*x.n.2.7*	WBS Dictionary	A document that describes each component in the work breakdown structure (WBS). For each WBS component, the WBS dictionary includes a brief definition of the scope or statement of work, defined deliverable(s), a list of associated activities, and a list of milestones. Other information may include: responsible organization, start and end dates, resources required, an estimate of cost, charge number, contract information, quality requirements, and technical references to facilitate performance of the work.
4	*x.n.2.8*	Risk Planning	*Added to summarize risk activities in Planning Process Group.*
5	*x.n.2.8.1*	Risk Register	The document containing the results of the qualitative risk analysis, quantitative risk analysis and risk response planning. The risk register details all identified risks, including description, category, cause, probability of occurring, impact(s) on objectives, proposed responses, owners, and current status. The risk register is a component of the project management plan.
5	*x.n.2.8.2*	Risk Related Contractual Agreements	

5	*x.n.2.8.3*	Risk Breakdown Structure	A hierarchically organized depiction of the identified project risks arranged by risk category and subcategory that identifies the various areas and causes of potential risks. The risk breakdown structure is often tailored to specific project types.
5	*x.n.2.8.4*	Probability and Impact Matrix	A common way to determine whether a risk is considered low, moderate or high by combining the two dimensions of a risk: its probability of occurrence, and its impact on objectives if it occurs.
5	*x.n.2.8.5*	Prioritized List of Quantified Risks	
5	*x.n.2.8.6*	Assumptions Analysis	A technique that explores the accuracy of assumptions and identifies risks to the project from inaccuracy, inconsistency, or incompleteness of assumptions.
5	*x.n.2.8.7*	Risk Control Procedures	
5	*x.n.2.8.8*	Risk Data Quality Assessment	
5	*x.n.2.8.9*	Sensitivity Analysis	A quantitative risk analysis and modeling technique used to help determine which risks have the most potential impact on the project. It examines the extent to which the uncertainty of each project element affects the objective being examined when all other uncertain elements are held at their baseline values. The typical display of results is in the form of a tornado diagram.

(continues)

Table C.1 *(continued)*

WBS LEVEL	WBS Code	WBS Element Title	Description (*PMBOK*® *Guide*—Third Edition Glossary or Applicable Errata)
1	***x***	Project Name	Entire project scope including all other project deliverables. Represents 100% of the project scope.
5	*x.n.2.8.10*	Expected Monetary Value (EMV) Analysis	A statistical technique that calculates the average outcome when the future includes scenarios that may or may not happen. A common use of this technique is within decision tree analysis. Modeling and simulation are recommended for cost and schedule risk analysis because it is more powerful and less subject to misapplication than expected monetary value analysis.
5	*x.n.2.8.11*	Decision Tree Analysis	The decision tree is a diagram that describes a decision under consideration and the implications of choosing one or another of the available alternatives. It is used when some future scenarios or outcomes of actions are uncertain. It incorporates probabilities and the costs or rewards of each logical path of events and future decisions, and uses expected monetary value analysis to help the organization identify the relative values of alternate actions.
5	*x.n.2.8.12*	Probabilistic Analysis	
5	*x.n.2.8.13*	Reserve Analysis	

4	*x.n.2.9*	Organization Charts and Position Descriptions	Organization Chart—A method for depicting interrelationships among a group of persons working together toward a common objective. Position Description—An explanation of a project team member's roles and responsibilities.
5	*x.n.2.9.1*	Roles and Responsibilities	
5	*x.n.2.9.2*	Project Organization Charts	A document that graphically depicts the project team members and their interrelationships for a specific project.
4	*x.n.2.10*	Project Management Plan	A formal, approved document that defines how the project is executed, monitored and controlled. It may be summary or detailed and may be composed of one or more subsidiary management plans and other planning documents.
5	*x.n.2.10.1*	Project Scope Management Plan	The document that describes how the project scope will be defined, developed, and verified and how the work breakdown structure will be created and defined, and that provides guidance on how the project scope will be managed and controlled by the project management team. It is contained in or is a subsidiary plan of the project management plan. The project scope management plan can be informal and broadly framed, or formal and highly detailed, based on the needs of the project.

(continues)

Table C.1 *(continued)*

WBS LEVEL	WBS Code	WBS Element Title	Description (*PMBOK® Guide*—Third Edition Glossary or Applicable Errata)
1	*x*	Project Name	Entire project scope including all other project deliverables. Represents 100% of the project scope.
5	*x.n.2.10.2*	Cost Management Plan	The document that sets out the format and establishes the activities and criteria for planning, structuring, and controlling the project costs. A cost management plan can be formal or informal, highly detailed or broadly framed, based on the requirements of the project stakeholders. The cost management plan is contained in, or is a subsidiary plan, of the project management plan.
5	*x.n.2.10.3*	Quality Management Plan	The quality management plan describes how the project management team will implement the performing organization's quality policy. The quality management plan is a component or a subsidiary plan of the project management plan. The quality management plan may be formal or informal, highly detailed, or broadly framed, based on the requirements of the project.
5	*x.n.2.10.4*	Staffing Management Plan	The document that describes when and how human resource requirements will be met. It is contained in, or is a subsidiary plan of, the project management plan. The staffing management plan can be informal and broadly framed, or formal and highly detailed, based on the needs of the project. Information in the staffing management plan varies by application area and project size.

5	*x.n.2.10.5*	Schedule Management Plan	The document that established criteria and the activities for developing and controlling the project schedule. It is contained in, or is a subsidiary plan of, the project management plan. The schedule management plan may be formal or informal, highly detailed or broadly framed, based on the needs of the project.
5	*x.n.2.10.6*	Communications Management Plan	The document that describes: the communication needs and expectations for the project; how and in what format information will be communicated; when and where each communication will be made; and who is responsible for providing each type of communication. A communication management plan can be formal or informal, highly detailed or broadly framed, based on the requirements of the project stakeholders. The communication management plan is contained in, or is a subsidiary plan of, the project management plan.
5	*x.n.2.10.7*	Risk Management Plan	The document describing how project risk management will be structured and performed on the project. It is contained in or is a subsidiary plan of the project management plan. The risk management plan can be informal and broadly framed, or formal and highly detailed, based on the needs of the project. Information in the risk management plan varies by application area and project size. The risk management plan is different from the risk register that contains the list of project risks, the results of risk analysis, and the risk responses.

(*continues*)

Table C.1 (*continued*)

WBS LEVEL	WBS Code	WBS Element Title	Description (*PMBOK*® *Guide*—Third Edition Glossary or Applicable Errata)
1	***x***	Project Name	Entire project scope including all other project deliverables. Represents 100% of the project scope.
5	*x.n.2.10.8*	Procurement Management Plan	The document that describes how procurement processes from developing procurement documentation through contract closure will be managed.
5	*x.n.2.10.9*	Contract Management Plan	The document that describes how a specific contract will be administered and can include items such as required documentation delivery and performance requirements. A contract management plan can be formal or informal, highly detailed or broadly framed, based on the requirements in the contract. Each contract management plan is a subsidiary plan of the project management plan.
4	*x.n.2.11*	Project Funding Requirements	
4	*x.n.2.12*	Quality Planning	*Added to summarize quality activities in Planning Process Group.*
5	*x.n.2.12.1*	Quality Metrics	
5	*x.n.2.12.2*	Quality Checklists	
4	*x.n.2.13*	Schedule Planning	*Added to summarize schedule activities in Planning Process Group.*
5	*x.n.2.13.1*	Activity List	A documented tabulation of schedule activities that shows the activity description, activity identifier, and a sufficiently detailed scope of work description so project team members understand what work is to be performed.

5	*x.n.2.13.2*	Activity Attributes	Multiple attributes associated with each schedule activity that can be included within the activity list. Activity attributes include activity codes, predecessor activities, successor activities, logical relationships, leads and lags, resource requirements, imposed dates, constraints and assumptions.
5	*x.n.2.13.3*	Milestone List	
5	*x.n.2.13.4*	Activity Duration Estimates	
5	*x.n.2.13.5*	Activity Cost Estimates	A quantitative assessment of the likely cost of the resources required to complete schedule activities.
5	*x.n.2.13.6*	Activity Resource Requirements	
5	*x.n.2.13.7*	Resource Breakdown Structure	A hierarchical structure of resources by resource category and resource type used in resource leveling schedules and to develop resource limited schedules, and which may be used to identify and analyze project human resource assignments.
5	*x.n.2.13.8*	Resource Calendars	A calendar of working days and nonworking days that determines those dates on which each specific resource is idle or can be active. Typically defines resource specific holidays and resource availability periods.
5	*x.n.2.13.9*	Project Calendar	A calendar of working days or shifts that establishes those dates on which schedule activities are worked and nonworking days that determine those dates on which schedule activities are idle. Typically defines holidays, weekends and shift hours.

(*continues*)

Table C.1 *(continued)*

WBS LEVEL	WBS Code	WBS Element Title	Description (*PMBOK*® *Guide*—Third Edition Glossary or Applicable Errata)
1	***x***	Project Name	Entire project scope including all other project deliverables. Represents 100% of the project scope.
5	*x.n.2.13.10*	Project Schedule Network Diagram	Any schematic display of the logical relationships among the project schedule activities. Always drawn from left to right to reflect project work chronology.
5	*x.n.2.13.11*	Schedule Model Data	Supporting data for the project schedule.
5	*x.n.2.13.12*	Project Schedule	The planned dates for performing schedule activities and the planned dates for meeting schedule milestones.
4	*x.n.2.14*	Procurement Planning	*Added to summarize procurement activities in Planning Process Group.*
5	*x.n.2.14.1*	Make or Buy Decisions	
5	*x.n.2.14.2*	Procurement Documents	Those documents utilized in bid and proposal activities, which include buyer's Invitation for Bid, Invitation for Negotiations, Request for Information, Request for Quotation, Request for Proposal and seller's responses.
5	*x.n.2.14.3*	Evaluation Criteria	
5	*x.n.2.14.4*	Selected Sellers	

4	*x.n.2.15*	Constraints	The state, quality, or sense of being restricted to a given course of action or inaction. An applicable restriction or limitation, either internal or external to the project, that will affect the performance of the project or a process. For example, a schedule constraint is any limitation or restraint placed on the project schedule that affects when a schedule activity can be scheduled and is usually in the form of fixed imposed dates. A cost constraint is any limitation or restraint placed on the project budget such as funds available over time. A project resource constraint is any limitation or restraint placed on resource usage, such as what resource skills or disciplines are available and the amount of a given resource available during a specified time frame.
4	*x.n.2.16*	Assumptions	Assumptions are factors that, for planning purposes, are considered to be true, real, or certain without proof or demonstration. Assumptions affect all aspects of project planning, and are part of the progressive elaboration of the project. Project teams frequently identify, document, and validate assumptions as part of their planning process. Assumptions generally involve a degree of risk.
4	*x.n.2.17*	Change Control System	A collection of formal documented procedures that define how project deliverables and documentation will be controlled, changed, and approved. In most application areas the change control system is a subset of the configuration management system.

(*continues*)

Table C.1 *(continued)*

WBS LEVEL	WBS Code	WBS Element Title	Description (*PMBOK*® *Guide*—Third Edition Glossary or Applicable Errata)
1	***x***	Project Name	Entire project scope including all other project deliverables. Represents 100% of the project scope.
4	*x.n.2.18*	Schedule Baseline	
4	*x.n.2.19*	Cost Baseline	
4	*x.n.2.20*	Quality Baseline	
4	*x.n.2.21*	Process Improvement Plan	
4	*x.n.2.22*	Communications Planning	*Added to summarize communications activities in Planning Process Group.*
5	*x.n.2.22.1*	Communications Requirements Analysis	
5	*x.n.2.22.2*	Lessons Learned	
5	*x.n.2.22.3*	Glossary of Common Terminology	
4	*x.n.2.23*	Change Requests	Requests to expand or reduce the project scope, modify policies, processes, plans, or procedures, modify costs or budgets, or revise schedules. Requests for a change can be direct or indirect, externally or internally initiated, and legally or contractually mandated or optional. Only formally documented requested changes are processed and only approved change requests are implemented.
5	*x.n.2.23.1*	Requested Changes	A formally documented change request that is submitted for approval to the integrated change control process. Contrast with approved change request.

5	*x.n.2.23.1*	Approved Change Requests	A change request that has been processed through the integrated change control process and approved. Contrast with requested change.
3	***x.n.3***	**Executing**	**This level represents the summary level for the Executing process group**.
4	*x.n.3.1*	Project Team Acquisition	
5	*x.n.3.1.1*	Project Staff Assignments	
5	*x.n.3.1.2*	Resource Availability	
4	*x.n.3.2*	Quality	*Added to summarize quality activities in Executing Process Group.*
5	*x.n.3.2.1*	Work Performance Information	Information and data, on the status of the project schedule activities being performed to accomplish the project work, collected as part of the direct and manage project execution processes. Information includes: status of deliverables; implementation status for change requests, corrective actions, preventative actions, and defect repairs; forecasted estimates to complete; reported percent of work physically completed; achieved value of technical performance measures; start and finish dates of schedule activities.
5	*x.n.3.2.2*	Quality Control Measurements	
5	*x.n.3.2.3*	Quality Audits	
4	*x.n.3.3*	Procurement	*Added to summarize procurement activities in Executing Process Group.*
5	*x.n.3.3.1*	Qualified Sellers List	

(continues)

Table C.1 *(continued)*

WBS LEVEL	WBS Code	WBS Element Title	Description (*PMBOK*® *Guide*—Third Edition Glossary or Applicable Errata)
1	***x***	Project Name	Entire project scope including all other project deliverables. Represents 100% of the project scope.
5	*x.n.3.3.2*	Bidder Conferences	
5	*x.n.3.3.3*	Advertising	
5	*x.n.3.3.4*	Procurement Document Package	
5	*x.n.3.3.5*	Proposals	
5	*x.n.3.3.7*	Screening System	
5	*x.n.3.3.8*	Contract Negotiation	
4	*x.n.3.4*	Changes	Impacts or potential impacts to the project that must be controlled.
5	*x.n.3.4.1*	Implemented Change Requests	
5	*x.n.3.4.2*	Corrective Actions	Documented direction for executing the project work to bring expected future performance of the project work in line with the project management plan.
6	*x.n.3.4.2.1*	Approved Corrective Actions	
6	*x.n.3.4.2.2*	Implemented Corective Actions	
5	*x.n.3.4.3*	Preventive Actions	Documented direction to perform an activity that can reduce the probability of negative consequences associated with project risks.
6	*x.n.3.4.3.1*	Approved Preventive Actions	

6	*x.n.3.4.3.2*	Implemented Preventive Actions	
5	*x.n.3.4.4*	Defect Repairs	Formally documented identification of a defect in a project component with a recommendation to either repair the defect or completely replace the component.
6	*x.n.3.4.4.1*	Approved Defect Repairs	
6	*x.n.3.4.4.2*	Implemented Defect Repairs	
6	*x.n.3.4.4.3*	Validated Defect Repairs	
4	*x.n.3.5*	Communications Deliverables	*Added to summarize communications activities in Executing Process Group.*
5	*x.n.3.5.1*	Project Presentations	
5	*x.n.3.5.2*	Team Performance Assessment	
4	*x.n.3.6*	Training	
4	*x.n.3.7*	Recognition and Rewards	
3	***x.n.4***	**Monitoring and Controlling**	**This level represents the summary level for the Monitoring and Controlling process group.**
4	*x.n.4.1*	Risk Audits	
4	*x.n.4.2*	Procurement Monitoring and Controlling	*Added to summarize procurement activities in Monitoring and Controlling Process Group.*
5	*x.n.4.2.1*	Seller Performance Evaluation Documentation	

(continues)

Table C.1 *(continued)*

WBS LEVEL	WBS Code	WBS Element Title	Description (*PMBOK® Guide*—Third Edition Glossary or Applicable Errata)
1	***x***	Project Name	Entire project scope including all other project deliverables. Represents 100% of the project scope.
5	*x.n.4.2.2*	Contract Documentation	
4	*x.n.4.3*	Project Reports	
5	*x.n.4.3.1*	Forecasts	Estimates or predictions of conditions and events in the project's future based on information and knowledge available at the time of the forecast. Forecasts are updated and reissued based on work performance information provided as the project is executed. The information is based on the project's past performance and expected future performance, and includes information that could impact the project in the future, such as estimate at completion and estimate to complete.
5	*x.n.4.3.2*	Issue Log	
5	*x.n.4.3.3*	Performance Reports	Documents and presentations that provide organized and summarized work performance information, earned value management parameters and calculations, and analyses of project work progress and status. Common formats for performance reports include bar charts, S-curves, histograms, tables and project schedule network diagram showing current schedule progress.

4	*x.n.4.4*	Performance Measurements	
4	*x.n.4.5*	Performance Measurement Baseline	An approved integrated scope-schedule-cost plan for the project work against which project execution is compared to measure and manage performance. Technical and quality parameters may also be included.
4	*x.n.4.6*	Project Performance Reviews	
4	*x.n.4.7*	Accepted Deliverables	
3	***x.n.5***	**Closing**	**This level represents the summary level for the Closing process group.**
4	*x.n.5.1*	Formal Acceptance Documentation	
4	*x.n.5.2*	Administrative Closure Procedure	
4	*x.n.5.3*	Contract File	
4	*x.n.5.4*	Procurement Audits	
4	*x.n.5.5*	Deliverable Acceptance	
4	*x.n.5.6*	Contract Closure Procedure	
4	*x.n.5.7*	Project Closure Documents	
4	*x.n.5.8*	Closed Contracts	

Source: Project Management Institute, A Guide to the Project Management Body of Knowledge (*PMBOK® Guide*—Third Edition) PMI. Newtown Square: PA.

Table C.2 Knowledge Area View

WBS Level	WBS Code	WBS Element Title	Description (*PMBOK*® *Guide*—Third Edition Glossary or Applicable Errata)	Process Group				
				Initiating	**Planning**	**Executing**	**Monitoring & Controlling**	**Closing**
1	***x***	**Project Name**	**Entire project scope including all other project deliverables. Represents 100% of the project scope.**					
2	***x.n***	**Project Management**						
3	***x.n.1***	**Project Integration Management**	**The processes and activities needed to identify, define, combine, unify, and coordinate the various processes and project management activities within the Project Management Process Groups.**					
4	*x.n.1.1*	Contract	A contract is a mutually binding agreement that obligates the seller to provide the specified product or service or result and obligates the buyer to pay for it.	x		x	x	x
4	*x.n.1.2*	Project Statement of Work		x				

4	*x.n.1.3*	Project Charter	A document issued by the project initiator or sponsor that formally authorizes the existence of a project, and provides the project manager with the authority to apply organizational resources to project activities.	x				
4	*x.n.1.4*	Project Management Plan	A formal, approved document that defines how the project is executed, monitored and controlled. It may be summary or detailed and may be composed of one or more subsidiary management plans and other planning documents.		x			
5	*x.n.1.4.1*	Project Scope Management Plan	The document that describes how the project scope will be defined, developed, and verified and how the work breakdown structure will be created and defined, and that provides guidance on how the project scope will be managed and controlled by the project management team. It is contained in or is a subsidiary plan of the project management plan. The project scope management plan can be informal and broadly framed, or formal and highly detailed, based on the needs of the project.		x			

(continues)

Table C.2 *(continued)*

WBS Level	WBS Code	WBS Element Title	Description (*PMBOK*® *Guide*—Third Edition Glossary or Applicable Errata)	Process Group				
				Initiating	**Planning**	**Executing**	**Monitoring & Controlling**	**Closing**
1	*x*	**Project Name**	**Entire project scope including all other project deliverables. Represents 100% of the project scope.**					
5	*x.n.1.4.2*	Cost Management Plan	The document that sets out the format and establishes the activities and criteria for planning, structuring, and controlling the project costs. A cost management plan can be formal or informal, highly detailed or broadly framed, based on the requirements of the project stakeholders. The cost management plan is contained in, or is a subsidiary plan, of the project management plan.		x			

5	*x.n.1.4.3*	Quality Management Plan	The quality management plan describes how the project management team will implement the performing organization's quality policy. The quality management plan is a component or a subsidiary plan of the project management plan. The quality management plan may be formal or informal, highly detailed, or broadly framed, based on the requirements of the project.		x			
5	*x.n.1.4.4*	Staffing Management Plan	The document that describes when and how human resource requirements will be met. It is contained in, or is a subsidiary plan of, the project management plan. The staffing management plan can be informal and broadly framed, or formal and highly detailed, based on the needs of the project. Information in the staffing management plan varies by application area and project size.		x			

(continues)

Table C.2 *(continued)*

WBS Level	WBS Code	WBS Element Title	Description (*PMBOK*® *Guide*—Third Edition Glossary or Applicable Errata)	Process Group				
1	*x*	**Project Name**	**Entire project scope including all other project deliverables. Represents 100% of the project scope**.	**Initiating**	**Planning**	**Executing**	**Monitoring & Controlling**	**Closing**
5	*x.n.1.4.5*	Schedule Management Plan	The document that established criteria and the activities for developing and controlling the project schedule. It is contained in, or is a subsidiary plan of, the project management plan. The schedule management plan may be formal or informal, highly detailed or broadly framed, based on the needs of the project.		x			

5	*x.n.1.4.6*	Communications Management Plan	The document that describes: the communication needs and expectations for the project; how and in what format information will be communicated; when and where each communication will be made; and who is responsible for providing each type of communication. A communication management plan can be formal or informal, highly detailed or broadly framed, based on the requirements of the project stakeholders. The communication management plan is contained in, or is a subsidiary plan of, the project management plan.		x			

(continues)

Table C.2 (*continued*)

WBS Level	WBS Code	WBS Element Title	Description (*PMBOK*® *Guide*—Third Edition Glossary or Applicable Errata)	Process Group				
1	*x*	**Project Name**	**Entire project scope including all other project deliverables. Represents 100% of the project scope**.	**Initiating**	**Planning**	**Executing**	**Monitoring & Controlling**	**Closing**
5	*x.n.1.4.7*	Risk Management Plan	The document describing how project risk management will be structured and performed on the project. It is contained in or is a subsidiary plan of the project management plan. The risk management plan can be informal and broadly framed, or formal and highly detailed, based on the needs of the project. Information in the risk management plan varies by application area and project size. The risk management plan is different from the risk register that contains the list of project risks, the results of risk analysis, and the risk responses.		x			

5	*x.n.1.4.8*	Procurement Management Plan	The document that describes how procurement processes from developing procurement documentation through contract closure will be managed.		x			
5	*x.n.1.4.9*	Contract Management Plan	The document that describes how a specific contract will be administered and can include items such as required documentation delivery and performance requirements. A contract management plan can be formal or informal, highly detailed or broadly framed, based on the requirements in the contract. Each contract management plan is a subsidiary plan of the project management plan.		x	x	x	
4	*x.n.1.5*	Changes	Impacts or potential impacts to the project that must be controlled.					

(continues)

Table C.2 (*continued*)

WBS Level	WBS Code	WBS Element Title	Description (*PMBOK*® *Guide*—Third Edition Glossary or Applicable Errata)	Process Group				
1	*x*	**Project Name**	**Entire project scope including all other project deliverables. Represents 100% of the project scope.**	**Initiating**	**Planning**	**Executing**	**Monitoring & Controlling**	**Closing**
5	*x.n.1.5.1*	Change Requests	Requests to expand or reduce the project scope, modify policies, processes, plans, or procedures, modify costs or budgets, or revise schedules. Requests for a change can be direct or indirect, externally or internally initiated, and legally or contractually mandated or optional. Only formally documented requested changes are processed and only approved change requests are implemented.					
6	*x.n.1.5.1.1*	Requested Changes	A formally documented change request that is submitted for approval to the integrated change control process. Contrast with approved change request.		x	x	x	

6	*x.n.1.5.1.2*	Approved Change Requests	A change request that has been processed through the integrated change control process and approved. Contrast with requested change.		x		x	
6	*x.n.1.5.1.3*	Implemented Change Requests				x		
5	*x.n.1.5.2*	Corrective Actions	Documented direction for executing the project work to bring expected future performance of the project work in line with the project management plan.					
6	*x.n.1.5.2.1*	Approved Corrective Actions				x	x	
6	*x.n.1.5.2.2*	Implemented Corrective Actions				x		
5	*x.n.1.5.3*	Preventive Actions	Documented direction to perform an activity that can reduce the probability of negative consequences associated with project risks.					
6	*x.n.1.5.3.1*	Approved Preventive Actions				x	x	

(continues)

Table C.2 *(continued)*

WBS Level	WBS Code	WBS Element Title	Description (*PMBOK*® *Guide*—Third Edition Glossary or Applicable Errata)	Process Group				
1	***x***	**Project Name**	**Entire project scope including all other project deliverables. Represents 100% of the project scope.**	**Initiating**	**Planning**	**Executing**	**Monitoring & Controlling**	**Closing**
6	*x.n.1.5.3.2*	Implemented Preventive Actions				x		
5	*x.n.1.5.4*	Defect Repairs	Formally documented identification of a defect in a project component with a recommendation to either repair the defect or completely replace the component.					
6	*x.n.1.5.4.1*	Approved Defect Repairs				x	x	
6	*x.n.1.5.4.2*	Implemented Defect Repairs				x		
6	*x.n.1.5.4.3*	Validated Defect Repairs				x	x	
4	*x.n.1.6*	Formal Acceptance Documentation						x

4	*x.n.1.7*	Project Closure Documents						x
3	***x.n.2***	**Project Scope Management**	**The processes required to ensure that the project includes all the work required, and only the work required, to complete the project successfully**.					
4	*x.n.2.1*	Preliminary Project Scope Statement			x			
4	*x.n.2.2*	Project Scope Statement	The narrative description of the project scope, including major deliverables, projects objectives, project assumptions, project constraints, and a statement of work, that provides a documented basis for making future project decisions and for confirming or developing a common understanding of project scope among the stakeholders. The definition of the project scope—what needs to be accomplished.		x			

(continues)

Table C.2 (*continued*)

WBS Level	WBS Code	WBS Element Title	Description (*PMBOK® Guide*—Third Edition Glossary or Applicable Errata)	Process Group				
1	*x*	**Project Name**	**Entire project scope including all other project deliverables. Represents 100% of the project scope.**	**Initiating**	**Planning**	**Executing**	**Monitoring & Controlling**	**Closing**
4	*x.n.2.3*	Work Breakdown Structure	A deliverable-oriented hierarchical decomposition of the work to be executed by the project team to accomplish the project objectives and create the required deliverables. It organizes and defines the total scope of the project. Each descending level represents an increasingly detailed definition of the project work. The WBS is decomposed into work packages. The deliverable orientation of the hierarchy includes both internal and external deliverables.		x			

4	*x.n.2.4*	WBS Dictionary	A document that describes each component in the work breakdown structure (WBS). For each WBS component, the WBS dictionary includes a brief definition of the scope or statement of work, defined deliverable(s), a list of associated activities, and a list of milestones. Other information may include: responsible organization, start and end dates, resources required, an estimate of cost, charge number, contract information, quality requirements, and technical references to facilitate performance of the work.				x	
4	*x.n.2.5*	Contract Statement of Work	A narrative description of products, services, or results to be supplied under contract.		x			
4	*x.n.2.6*	Scope Baseline			x			

(continues)

Table C.2 *(continued)*

WBS Level	WBS Code	WBS Element Title	Description (*PMBOK® Guide*—Third Edition Glossary or Applicable Errata)	Process Group				
1	***x***	**Project Name**	**Entire project scope including all other project deliverables. Represents 100% of the project scope**.	**Initiating**	**Planning**	**Executing**	**Monitoring & Controlling**	**Closing**
4	*x.n.2.7*	Product Scope Description	The documented narrative description of the product scope.		x			
4	*x.n.2.8*	Accepted Deliverables					x	
3	***x.n.3***	**Project Time Management**	**The processes required to accomplish timely completion of the project**.					
4	*x.n.3.1*	Activity List	A documented tabulation of schedule activities that shows the activity description, activity identifier, and a sufficiently detailed scope of work description so project team members understand what work is to be performed.		x			

4	*x.n.3.2*	Activity Attributes	Multiple attributes associated with each schedule activity that can be included within the activity list. Activity attributes include activity codes, predecessor activities, successor activities, logical relationships, leads and lags, resource requirements, imposed dates, constraints and assumptions.		x			
4	*x.n.3.3*	Milestone List			x			
4	*x.n.3.4*	Activity Duration Estimates			x			
4	*x.n.3.5*	Activity Resource Requirements			x			
4	*x.n.3.6*	Resource Calendar	A calendar of working days and nonworking days that determines those dates on which each specific resource is idle or can be active. Typically defines resource specific holidays and resource availability periods.		x			

(continues)

Table C.2 *(continued)*

WBS Level	WBS Code	WBS Element Title	Description (*PMBOK*® *Guide*—Third Edition Glossary or Applicable Errata)	Process Group				
1	*x*	**Project Name**	**Entire project scope including all other project deliverables. Represents 100% of the project scope.**	**Initiating**	**Planning**	**Executing**	**Monitoring & Controlling**	**Closing**
4	*x.n.3.7*	Project Calendar	A calendar of working days or shifts that establishes those dates on which schedule activities are worked and nonworking days that determine those dates on which schedule activities are idle. Typically defines holidays, weekends and shift hours.		x			
4	*x.n.3.8*	Project Schedule Network Diagram	Any schematic display of the logical relationships among the project schedule activities. Always drawn from left to right to reflect project work chronology.		x			
4	*x.n.3.9*	Schedule Model Data			x			
4	*x.n.3.10*	Project Schedule	The planned dates for performing schedule activities and the planned dates for meeting schedule milestones.		x			

4	*x.n.3.11*	Schedule Baseline			x			
3	***x.n.4***	**Project Cost Management**	**The processes involved in planning, estimating, budgeting, and controlling costs so that the project can be completed within the approved budget.**					
4	*x.n.4.1*	Activity Cost Estimates			x			
4	*x.n.4.2*	Project Funding Requirements			x			
4	*x.n.4.3*	Cost Baseline					x	
4	*x.n.4.4*	Project Performance Reviews					x	
3	***x.n.5***	**Project Quality Management**	**The processes and activities of the performing organization that determine quality policies, objectives, and responsibilities so that the project will satisfy the needs for which it was undertaken.**					
4	*x.n.5.1*	Quality Metrics			x			
4	*x.n.5.2*	Quality Checklists			x			
4	*x.n.5.3*	Quality Baseline			x			

(continues)

Table C.2 (*continued*)

WBS Level	WBS Code	WBS Element Title	Description (*PMBOK*® *Guide*—Third Edition Glossary or Applicable Errata)	Process Group				
				Initiating	Planning	Executing	Monitoring & Controlling	Closing
1	***x***	**Project Name**	**Entire project scope including all other project deliverables. Represents 100% of the project scope.**					
4	*x.n.5.4*	Quality Control Measurements				x	x	
4	*x.n.5.5*	Quality Audits				x		
3	***x.n.6***	**Project Human Resources Management**	**The processes that organize and manage the project team.**					
4	*x.n.6.1*	Organization Charts and Position Descriptions	Organization Chart—A method for depicting interrelationships among a group of persons working together toward a common objective. Position Description—An explanation of a project team member's roles and responsibilities.		x			
5	*x.n.6.1.1*	Roles and Responsibilities			x			

5	*x.n.6.1.2*	Project Organization Charts	A document that graphically depicts the project team members and their interrelationships for a specific project.		x			
4	*x.n.6.2*	Project Team Acquisition				x		
5	*x.n.6.2.1*	Project Staff Assignments				x		
5	*x.n.6.2.2*	Resource Availability				x		
4	*x.n.6.3*	Team Performance Assessment				x		
4	*x.n.6.4*	Training				x		
4	*x.n.6.5*	Recognition and Rewards				x		
3	***x.n.7***	**Project Communication Management**	**The processes required to ensure timely and appropriate generation, collection, distribution, storage, retrieval and ultimate disposition of project information.**					

(continues)

Table C.2 *(continued)*

WBS Level	WBS Code	WBS Element Title	Description (*PMBOK*® *Guide*—Third Edition Glossary or Applicable Errata)	Process Group				
1	***x***	**Project Name**	**Entire project scope including all other project deliverables. Represents 100% of the project scope**.	**Initiating**	**Planning**	**Executing**	**Monitoring & Controlling**	**Closing**
4	*x.n.7.1*	Constraints	The state, quality, or sense of being restricted to a given course of action or inaction. An applicable restriction or limitation, either internal or external to the project, that will affect the performance of the project or a process. For example, a schedule constraint is any limitation or restraint placed on the project schedule that affects when a schedule activity can be scheduled and is usually in the form of fixed imposed dates. A cost constraint is any limitation or restraint placed on the project budget such as funds available over time. A project resource constraint is any limitation or restraint placed on resource usage, such as what resource skills or disciplines are available and the amount of a given resource available during a specified time frame.		x			

4	*x.n.7.2*	Assumptions	Assumptions are factors that, for planning purposes, are considered to be true, real, or certain without proof or demonstration. Assumptions affect all aspects of project planning, and are part of the progressive elaboration of the project. Project teams frequently identify, document, and validate assumptions as part of their planning process. Assumptions generally involve a degree of risk.		x			
4	*x.n.7.3*	Project Reports				x		
5	*x.n.7.3.1*	Forecasts	Estimates or predictions of conditions and events in the project's future based on information and knowledge available at the time of the forecast. Forecasts are updated and reissued based on work performance information provided as the project is executed. The information is based on the project's past performance and expected future performance, and includes information that could impact the project in the future, such as estimate at completion and estimate to complete.				x	

(continues)

Table C.2 (*continued*)

WBS Level	WBS Code	WBS Element Title	Description (*PMBOK*® *Guide*—Third Edition Glossary or Applicable Errata)	Process Group				
1	***x***	**Project Name**	**Entire project scope including all other project deliverables. Represents 100% of the project scope.**	**Initiating**	**Planning**	**Executing**	**Monitoring & Controlling**	**Closing**
5	*x.n.7.3.2*	Resource Breakdown Structure	A hierarchical structure of resources by resource category and resource type used in resource leveling schedules and to develop resource limited schedules, and which may be used to identify and analyze project human resource assignments.		x			
5	*x.n.7.3.3*	Issue Log					x	
4	*x.n.7.4*	Performance Measurement Baseline	An approved integrated scope-schedule-cost plan for the project work against which project execution is compared to measure and manage performance. Technical and quality parameters may also be included.				x	

4	*x.n.7.5*	Work Performance Information	Information and data, on the status of the project schedule activities being performed to accomplish the project work, collected as part of the direct and manage project execution processes. Information includes: status of deliverables; implementation status for change requests, corrective actions, preventative actions, and defect repairs; forecasted estimates to complete; reported percent of work physically completed; achieved value of technical performance measures; start and finish dates of schedule activities.			x		
4	*x.n.7.6*	Performance Measurements					x	
4	*x.n.7.7*	Performance Reports	Documents and presentations that provide organized and summarized work performance information, earned value management parameters and calculations, and analyses of project work progress and status. Common formats for performance reports include bar charts, S-curves, histograms, tables and project schedule network diagram showing current schedule progress.				x	

(*continues*)

Table C.2 *(continued)*

WBS Level	WBS Code	WBS Element Title	Description (*PMBOK® Guide*—Third Edition Glossary or Applicable Errata)	Process Group				
1	***x***	**Project Name**	**Entire project scope including all other project deliverables. Represents 100% of the project scope.**	**Initiating**	**Planning**	**Executing**	**Monitoring & Controlling**	**Closing**
4	*x.n.7.8*	Organizational Process Assets	Any or all process related assets, from any or all of the organizations involved in the project that are or can be used to influence the project's success. The process assets include formal and informal plans, policies, procedures, and guidelines. The process assets also include the organization's knowledge bases such as lessons learned and historical information.	x	x	x	x	x
5	*x.n.7.8.1*	Administrative Closure Procedure						x
5	*x.n.7.8.2*	Contract Closure Procedure			x		x	x

5	*x.n.7.8.3*	Change Control System	A collection of formal documented procedures that define how project deliverables and documentation will be controlled, changed, and approved. In most application areas the change control system is a subset of the configuration management system.		x		x	
5	*x.n.7.8.4*	Risk Control Procedures			x			
5	*x.n.7.8.5*	Process Improvement Plan			x			
4	*x.n.7.9*	Other Communications Deliverables						
5	*x.n.7.9.1*	Communications Requirements Analysis			x			
5	*x.n.7.9.2*	Project Presentations				x		
5	*x.n.7.9.3*	Lessons Learned	The learning gained from the process of performing the project. Lessons learned may be identified at any point. Also considered a project record, to be included in the lessons learned knowledge base.		x	x	x	x

(continues)

Table C.2 (*continued*)

WBS Level	WBS Code	WBS Element Title	Description (*PMBOK*® *Guide*—Third Edition Glossary or Applicable Errata)	Process Group				
				Initiating	**Planning**	**Executing**	**Monitoring & Controlling**	**Closing**
1	***x***	**Project Name**	**Entire project scope including all other project deliverables. Represents 100% of the project scope.**					
5	*x.n.7.9.4*	Glossary of Common Terminology			x			
3	***x.n.8***	**Project Risk Management**	**The processes concerned with conducting risk management planning, identification, analysis, responses and monitoring and control on a project.**					
4	*x.n.8.1*	Risk Register	The document containing the results of the qualitative risk analysis, quantitative risk analysis and risk response planning. The risk register details all identified risks, including description, category, cause, probability of occurring, impact(s) on objectives, proposed responses, owners, and current status. The risk register is a component of the project management plan.		x			

4	*x.n.8.2*	Risk-Related Contractual Agreements			x			
4	*x.n.8.3*	Risk Breakdown Structure	A hierarchically organized depiction of the identified project risks arranged by risk category and subcategory that identifies the various areas and causes of potential risks. The risk breakdown structure is often tailored to specific project types.		x			
4	*x.n.8.4*	Probability and Impact Matrix	A common way to determine whether a risk is considered low, moderate or high by combining the two dimensions of a risk: its probability of occurrence, and its impact on objectives if it occurs.		x			
4	*x.n.8.5*	Assumptions Analysis	A technique that explores the accuracy of assumptions and identifies risks to the project from inaccuracy, inconsistency, or incompleteness of assumptions.		x			
4	*x.n.8.6*	Risk Data Quality Assessment			x			

(continues)

Table C.2 *(continued)*

WBS Level	WBS Code	WBS Element Title	Description (*PMBOK*® *Guide*—Third Edition Glossary or Applicable Errata)	Process Group				
1	*x*	**Project Name**	**Entire project scope including all other project deliverables. Represents 100% of the project scope.**	**Initiating**	**Planning**	**Executing**	**Monitoring & Controlling**	**Closing**
4	*x.n.8.7*	Sensitivity Analysis	A quantitative risk analysis and modeling technique used to help determine which risks have the most potential impact on the project. It examines the extent to which the uncertainty of each project element affects the objective being examined when all other uncertain elements are held at their baseline values. The typical display of results is in the form of a tornado diagram.		x			

4	*x.n.8.8*	Expected Monetary Value (EMV) Analysis	A statistical technique that calculates the average outcome when the future includes scenarios that may or may not happen. A common use of this technique is within decision tree analysis. Modeling and simulation are recommended for cost and schedule risk analysis because it is more powerful and less subject to misapplication than expected monetary value analysis.		x			
4	*x.n.8.9*	Decision Tree Analysis	The decision tree is a diagram that describes a decision under consideration and the implications of choosing one or another of the available alternatives. It is used when some future scenarios or outcomes of actions are uncertain. It incorporates probabilities and the costs or rewards of each logical path of events and future decisions, and uses expected monetary value analysis to help the organization identify the relative values of alternate actions.		x			

(continues)

Table C.2 *(continued)*

WBS Level	WBS Code	WBS Element Title	Description (*PMBOK® Guide*—Third Edition Glossary or Applicable Errata)	Process Group				
1	***x***	**Project Name**	**Entire project scope including all other project deliverables. Represents 100% of the project scope.**	**Initiating**	**Planning**	**Executing**	**Monitoring & Controlling**	**Closing**
4	*x.n.8.10*	Probabilistic Analysis			x			
4	*x.n.8.11*	Prioritized List of Quantified Risks			x			
4	*x.n.8.12*	Risk Audits					x	
4	*x.n.8.13*	Reserve Analysis	An analytical technique to determine the essential features and relationships of components in the project management plan to establish a reserve for the schedule duration, budget, estimated cost, or funds for a project.		x		x	
3	***x.n.9***	**Project Procurement Management**	**The processes to purchase or acquire the products, services, or results needed from outside the project team to perform the work.**					

4	*x.n.9.1*	Make or Buy Decisions			x			
4	*x.n.9.2*	Procurement Documents	Those documents utilized in bid and proposal activities, which include buyer's Invitation for Bid, Invitation for Negotiations, Request for Information, Request for Quotation, Request for Proposal and seller's responses.		x			
4	*x.n.9.3*	Qualified Sellers List				x		
4	*x.n.9.4*	Evaluation Criteria			x	x		
4	*x.n.9.5*	Bidder Conferences				x		
4	*x.n.9.6*	Advertising				x		
4	*x.n.9.7*	Procurement Document Package				x		
4	*x.n.9.8*	Proposals				x		
4	*x.n.9.9*	Selected Sellers			x	x	x	
4	*x.n.9.10*	Contract Documentation					x	

(continues)

Table C.2 (*continued*)

WBS Level	WBS Code	WBS Element Title	Description (*PMBOK®* *Guide*—Third Edition Glossary or Applicable Errata)	Process Group				
1	***x***	**Project Name**	**Entire project scope including all other project deliverables. Represents 100% of the project scope.**	**Initiating**	**Planning**	**Executing**	**Monitoring & Controlling**	**Closing**
4	*x.n.9.11*	Closed Contracts						x
4	*x.n.9.12*	Seller Performance Evaluation Documentation					x	
4	*x.n.9.13*	Screening System				x		
4	*x.n.9.14*	Contract Negotiation				x		
4	*x.n.9.15*	Procurement Audits						x
4	*x.n.9.16*	Contract File						x
4	*x.n.9.17*	Deliverable Acceptance						x

Source: Project Management Institute, A Guide to the Project Management Body of Knowledge (*PMBOK®* *Guide*—Third Edition) PMI. Newtown Square: PA.

Table C.3 Knowledge Area "Lite" View

WBS Level	WBS Code	WBS Element Title	Description (*PMBOK*® *Guide*—Third Edition Glossary or Applicable Errata)
1	***x***	**Project Name**	**Entire project scope including all other project deliverables. Represents 100% of the project scope.**
2	*x.n*	**Project Management**	
3	***x.n.1***	**Project Integration Management**	**The processes and activities needed to identify, define, combine, unify, and coordinate the various processes and project management activities within the Project Management Process Groups.**
4	*x.n.1.1*	Contract	A contract is a mutually binding agreement that obligates the seller to provide the specified product or service or result and obligates the buyer to pay for it.
4	*x.n.1.2*	Project Statement of Work	
4	*x.n.1.3*	Project Charter	A document issued by the project initiator or sponsor that formally authorizes the existence of a project, and provides the project manager with the authority to apply organizational resources to project activities.
4	*x.n.1.4*	Project Management Plan	A formal, approved document that defines how the project is executed, monitored and controlled. It may be summary or detailed and may be composed of one or more subsidiary management plans and other planning documents.

(continues)

Table C.3 *(continued)*

WBS Level	WBS Code	WBS Element Title	Description (*PMBOK*® *Guide*—Third Edition Glossary or Applicable Errata)
1	***x***	**Project Name**	**Entire project scope including all other project deliverables. Represents 100% of the project scope.**
4	*x.n.1.5*	Changes	Impacts or potential impacts to the project that must be controlled.
5	*x.n.1.5.1*	Change Requests	Requests to expand or reduce the project scope, modify policies, processes, plans, or procedures, modify costs or budgets, or revise schedules. Requests for a change can be direct or indirect, externally or internally initiated, and legally or contractually mandated or optional. Only formally documented requested changes are processed and only approved change requests are implemented.
5	*x.n.1.5.2*	Corrective Actions	Documented direction for executing the project work to bring expected future performance of the project work in line with the project management plan.
5	*x.n.1.5.3*	Preventive Actions	Documented direction to perform an activity that can reduce the probability of negative consequences associated with project risks.
5	*x.n.1.5.4*	Defect Repairs	Formally documented identification of a defect in a project component with a recommendation to either repair the defect or completely replace the component.

4	*x.n.1.6*	Formal Acceptance Documentation	
4	*x.n.1.7*	Project Closure Documents	
3	***x.n.2***	**Project Scope Management**	**The processes required to ensure that the project includes all the work required, and only the work required, to complete the project successfully**.
4	*x.n.2.1*	Preliminary Project Scope Statement	
4	*x.n.2.2*	Project Scope Statement	The narrative description of the project scope, including major deliverables, projects objectives, project assumptions, project constraints, and a statement of work, that provides a documented basis for making future project decisions and for confirming or developing a common understanding of project scope among the stakeholders. The definition of the project scope—what needs to be accomplished.
4	*x.n.2.3*	Work Breakdown Structure	A deliverable-oriented hierarchical decomposition of the work to be executed by the project team to accomplish the project objectives and create the required deliverables. It organizes and defines the total scope of the project. Each descending level represents an increasingly detailed definition of the project work. The WBS is decomposed into work packages. The deliverable orientation of the hierarchy includes both internal and external deliverables.

(*continues*)

Table C.3 (*continued*)

WBS Level	WBS Code	WBS Element Title	Description (*PMBOK*® *Guide*—Third Edition Glossary or Applicable Errata)
1	***x***	**Project Name**	**Entire project scope including all other project deliverables. Represents 100% of the project scope.**
4	*x.n.2.4*	WBS Dictionary	A document that describes each component in the work breakdown structure (WBS). For each WBS component, the WBS dictionary includes a brief definition of the scope or statement of work, defined deliverable(s), a list of associated activities, and a list of milestones. Other information may include: responsible organization, start and end dates, resources required, an estimate of cost, charge number, contract information, quality requirements, and technical references to facilitate performance of the work.
4	*x.n.2.5*	Contract Statement of Work	A narrative description of products, services, or results to be supplied under contract.
4	*x.n.2.6*	Scope Baseline	
4	*x.n.2.7*	Product Scope Description	The documented narrative description of the product scope.
4	*x.n.2.8*	Accepted Deliverables	

3	***x.n.3***	**Project Time Management**	**The processes required to accomplish timely completion of the project.**
4	*x.n.3.1*	Activity List	A documented tabulation of schedule activities that shows the activity description, activity identifier, and a sufficiently detailed scope of work description so project team members understand what work is to be performed.
4	*x.n.3.2*	Activity Attributes	Multiple attributes associated with each schedule activity that can be included within the activity list. Activity attributes include activity codes, predecessor activities, successor activities, logical relationships, leads and lags, resource requirements, imposed dates, constraints and assumptions.
4	*x.n.3.3*	Milestone List	
4	*x.n.3.4*	Activity Duration Estimates	
4	*x.n.3.5*	Activity Resource Requirements	
4	*x.n.3.6*	Resource Calendar	A calendar of working days and nonworking days that determines those dates on which each specific resource is idle or can be active. Typically defines resource specific holidays and resource availability periods.
4	*x.n.3.7*	Project Calendar	A calendar of working days or shifts that establishes those dates on which schedule activities are worked and nonworking days that determine those dates on which schedule activities are idle. Typically defines holidays, weekends and shift hours.

(*continues*)

Table C.3 *(continued)*

WBS Level	WBS Code	WBS Element Title	Description (*PMBOK*® *Guide*—Third Edition Glossary or Applicable Errata)
1	***x***	**Project Name**	**Entire project scope including all other project deliverables. Represents 100% of the project scope.**
4	*x.n.3.8*	Project Schedule Network Diagram	Any schematic display of the logical relationships among the project schedule activities. Always drawn from left to right to reflect project work chronology.
4	*x.n.3.9*	Schedule Model Data	
4	*x.n.3.10*	Project Schedule	The planned dates for performing schedule activities and the planned dates for meeting schedule milestones.
4	*x.n.3.11*	Schedule Baseline	
3	***x.n.4***	**Project Cost Management**	**The processes involved in planning, estimating, budgeting, and controlling costs so that the project can be completed within the approved budget.**
4	*x.n.4.1*	Activity Cost Estimates	
4	*x.n.4.2*	Project Funding Requirements	
4	*x.n.4.3*	Cost Baseline	
4	*x.n.4.4*	Project Performance Reviews	
3	***x.n.5***	**Project Quality Management**	**The processes and activities of the performing organization that determine quality policies, objectives, and responsibilities so that the project will satisfy the needs for which it was undertaken.**

4	*x.n.5.1*	Quality Metrics	
4	*x.n.5.2*	Quality Checklists	
4	*x.n.5.3*	Quality Baseline	
4	*x.n.5.4*	Quality Control Measurements	
4	*x.n.5.5*	Quality Audits	
3	***x.n.6***	**Project Human Resources Management**	**The processes that organize and manage the project team**.
4	*x.n.6.1*	Organization Charts and Position Descriptions	Organization Chart—A method for depicting interrelationships among a group of persons working together toward a common objective. Position Description—An explanation of a project team member's roles and responsibilities.
5	*x.n.6.1.1*	Roles and Responsibilities	
5	*x.n.6.1.2*	Project Organization Charts	A document that graphically depicts the project team members and their interrelationships for a specific project.
4	*x.n.6.2*	Project Team Acquisition	
5	*x.n.6.2.1*	Project Staff Assignments	
5	*x.n.6.2.2*	Resource Availability	
4	*x.n.6.3*	Team Performance Assessment	
4	*x.n.6.4*	Training	

(*continues*)

Table C.3 *(continued)*

WBS Level	WBS Code	WBS Element Title	Description (*PMBOK*® *Guide*—Third Edition Glossary or Applicable Errata)
1	***x***	**Project Name**	**Entire project scope including all other project deliverables. Represents 100% of the project scope**.
4	*x.n.6.5*	Recognition and Rewards	
3	***x.n.7***	**Project Communication Management**	**The processes required to ensure timely and appropriate generation, collection, distribution, storage, retrieval and ultimate disposition of project information**.
4	*x.n.7.1*	Constraints	The state, quality, or sense of being restricted to a given course of action or inaction. An applicable restriction or limitation, either internal or external to the project, that will affect the performance of the project or a process. For example, a schedule constraint is any limitation or restraint placed on the project schedule that affects when a schedule activity can be scheduled and is usually in the form of fixed imposed dates. A cost constraint is any limitation or restraint placed on the project budget such as funds available over time. A project resource constraint is any limitation or restraint placed on resource usage, such as what resource skills or disciplines are available and the amount of a given resource available during a specified time frame.

4	*x.n.7.2*	Assumptions	Assumptions are factors that, for planning purposes, are considered to be true, real, or certain without proof or demonstration. Assumptions affect all aspects of project planning, and are part of the progressive elaboration of the project. Project teams frequently identify, document, and validate assumptions as part of their planning process. Assumptions generally involve a degree of risk.
4	*x.n.7.3*	Project Reports	
5	*x.n.7.3.1*	Forecasts	Estimates or predictions of conditions and events in the project's future based on information and knowledge available at the time of the forecast. Forecasts are updated and reissued based on work performance information provided as the project is executed. The information is based on the project's past performance and expected future performance, and includes information that could impact the project in the future, such as estimate at completion and estimate to complete.
5	*x.n.7.3.2*	Resource Breakdown Structure	A hierarchical structure of resources by resource category and resource type used in resource leveling schedules and to develop resource limited schedules, and which may be used to identify and analyze project human resource assignments.
5	*x.n.7.3.3*	Issue Log	
4	*x.n.7.4*	Performance Measurement Baseline	An approved integrated scope-schedule-cost plan for the project work against which project execution is compared to measure and manage performance. Technical and quality parameters may also be included.

(continues)

Table C.3 *(continued)*

WBS Level	WBS Code	WBS Element Title	Description (*PMBOK*® *Guide*—Third Edition Glossary or Applicable Errata)
1	*x*	**Project Name**	**Entire project scope including all other project deliverables. Represents 100% of the project scope**.
4	*x.n.7.5*	Work Performance Information	Information and data, on the status of the project schedule activities being performed to accomplish the project work, collected as part of the direct and manage project execution processes. Information includes: status of deliverables; implementation status for change requests, corrective actions, preventative actions, and defect repairs; forecasted estimates to complete; reported percent of work physically completed; achieved value of technical performance measures; start and finish dates of schedule activities.
4	*x.n.7.6*	Performance Measurements	
4	*x.n.7.7*	Performance Reports	Documents and presentations that provide organized and summarized work performance information, earned value management parameters and calculations, and analyses of project work progress and status. Common formats for performance reports include bar charts, S-curves, histograms, tables and project schedule network diagram showing current schedule progress.
4	*x.n.7.8*	Organizational Process Assets	Any or all process related assets, from any or all of the organizations involved in the project that are or can be used to influence the project's success. The process assets include formal and informal plans, policies, procedures, and guidelines. The process assets also include the organization's knowledge bases such as lessons learned and historical information.
4	*x.n.7.9*	Other Communications Deliverables	

5	*x.n.7.9.1*	Project Presentations	
5	*x.n.7.9.2*	Lessons Learned	The learning gained from the process of performing the project. Lessons learned may be identified at any point. Also considered a project record, to be included in the lessons learned knowledge base.
3	***x.n.8***	**Project Risk Management**	**The processes concerned with conducting risk management planning, identification, analysis, responses and monitoring and control on a project.**
4	*x.n.8.1*	Risk Register	The document containing the results of the qualitative risk analysis, quantitative risk analysis and risk response planning. The risk register details all identified risks, including description, category, cause, probability of occurring, impact(s) on objectives, proposed responses, owners, and current status. The risk register is a component of the project management plan.
4	*x.n.8.2*	Risk-Related Contractual Agreements	
4	*x.n.8.3*	Risk Breakdown Structure	A hierarchically organized depiction of the identified project risks arranged by risk category and subcategory that identifies the various areas and causes of potential risks. The risk breakdown structure is often tailored to specific project types.
4	*x.n.8.4*	Probability and Impact Matrix	A common way to determine whether a risk is considered low, moderate or high by combining the two dimensions of a risk: its probability of occurrence, and its impact on objectives if it occurs.
4	*x.n.8.5*	Assumptions Analysis	A technique that explores the accuracy of assumptions and identifies risks to the project from inaccuracy, inconsistency, or incompleteness of assumptions.
4	*x.n.8.6*	Prioritized List of Quantified Risks	

(*continues*)

Table C.3 *(continued)*

WBS Level	WBS Code	WBS Element Title	Description (*PMBOK*® *Guide*—Third Edition Glossary or Applicable Errata)
1	***x***	**Project Name**	**Entire project scope including all other project deliverables. Represents 100% of the project scope.**
4	*x.n.8.7*	Risk Audits	
3	***x.n.9***	**Project Procurement Management**	**The processes to purchase or acquire the products, services, or results needed from outside the project team to perform the work.**
4	*x.n.9.1*	Make or Buy Decisions	
4	*x.n.9.2*	Procurement Documents	Those documents utilized in bid and proposal activities, which include buyer's Invitation for Bid, Invitation for Negotiations, Request for Information, Request for Quotation, Request for Proposal and seller's responses.
4	*x.n.9.3*	Evaluation Criteria	
4	*x.n.9.4*	Procurement Document Package	
4	*x.n.9.5*	Proposals	
4	*x.n.9.6*	Contract Documentation	
4	*x.n.9.7*	Closed Contracts	
4	*x.n.9.8*	Seller Performance Evaluation Documentation	
4	*x.n.9.9*	Contract File	
4	*x.n.9.10*	Deliverable Acceptance	

Source: Project Management Institute, A Guide to the Project Management Body of Knowledge (*PMBOK*® *Guide*—Third Edition) PMI. Newtown Square: PA.

REFERENCE

Project Management Institute. 2004. *A Guide to the Project Management Body of Knowledge (PMBOK® Guide—Third Edition)*. Newtown Square, PA: Project Management Institute.

Appendix D

Answers to Chapter Questions

This appendix details the questions and answers to the quiz at the end of each chapter. Also provided is a reference back to the point in the chapter where the answer can be found.

CHAPTER 1 QUESTIONS

1. According to current PMI standards, Work Breakdown Structures are:
   a. Task-oriented
   b. Process-oriented
   c. Deliverable-oriented
   d. Time-oriented

   Correct Answer

   c. Deliverable-oriented

   Answer Reference

   Chapter: 1—Background and Key Concepts
   Section: Defining Work Breakdown Structure

2. The elements at lowest level of the WBS are called ____________?
   a. Control Accounts
   b. Work Packages
   c. WBS Deliverables
   d. Lowest Level WBS Elements

   Correct Answer

   b. Work Packages

Reference

Chapter: 1—Background and Key Concepts
Section: Defining Work Breakdown Structure

3. The ________________ is utilized as the starting point for creating a WBS.
   a. Preliminary Project Scope Statement
   b. Product Scope Description
   c. Final Product Scope Statement
   d. Project Charter

Correct Answer

d. Project Charter

Reference

Chapter: 1—Background and Key Concepts
Section: A Brief Story as an Illustration

4. Which of the following are key characteristics of high quality Work Breakdown Structures?
   (Select all that Apply)
   a. Task-oriented
   b. Deliverable-oriented
   c. Hierarchical
   d. Includes only the end products, services or results of the project
   e. Completely applies the 100% Rule

Correct Answer

b. Deliverable-oriented
c. Hierarchical
e. Completely applies the 100% Rule

Reference

Chapter: 1—Background and Key Concepts
Section: Defining the WBS

5. Who initially developed Work Breakdown Structures?
   a. U.S. Department of Defense and NASA
   b. Builders of the Great Pyramids of Egypt
   c. Architects of the Roman Coliseum
   d. Russian Space Agency

Correct Answers

a. U.S. Department of Defense and NASA

Reference

Chapter: 1—Background and Key Concepts
Section: The Role of the Work Breakdown Structure

CHAPTER 2 QUESTIONS

1. Which of the following are Core Characteristics of a quality WBS? (Select all that Apply)
   a. Deliverable-oriented
   b. Task-oriented
   c. Hierarchical
   d. Includes only the end products, services or results of the project
   e. Uses nouns, verbs and adjectives
   f. Is created by those performing the work

Correct Answer

a. Deliverable-oriented
c. Hierarchical
g. Is created by those performing the work

Answer Reference

Chapter: 2—Applying WBS Attributes and Concepts
Section: WBS Core Characteristics

2. Which of the following is true for quality Work Breakdown Structures?
   a. Program/Project Management can occur at any level of the WBS
   b. Contain at least three levels of decomposition
   c. Clearly communicates project scope to all stakeholders
   d. Does not include a WBS Dictionary

Correct Answer

c. Clearly communicates project scope to all stakeholders

Answer Reference

Chapter: 2—Applying WBS Attributes and Concepts
Section: WBS Core Characteristics

3. Which of the following is true for WBS Use-Related Characteristics?
   a. Characteristics are consistent from project to project
   b. WBS quality depends on how well the specific content and types of elements address the full set of needs of the project
   c. Contain only discrete WBS elements
   d. Must be decomposed at least three levels

Correct Answer

b. WBS quality depends on how well the specific content and types of elements address the full set of needs of the project

Answer Reference

Chapter: 2—Applying WBS Attributes and Concepts
Section: WBS Use-Related Characteristics

4. Which of the following statements is true for any WBS? (Select all that Apply)
   a. WBS quality characteristics apply at all levels of scope definition
   b. Valid WBS representations only include graphical and outline views.
   c. Using a project management scheduling tool for WBS creation is helpful in differentiating between WBS elements and Project Schedule tasks and activities.
   d. Tools for creating and managing WBS are very mature and easy to use.

Correct Answer

a. WBS quality characteristics apply at all levels of scope definition

Answer Reference

Chapter: 2—Applying WBS Attributes and Concepts
Section: WBS Uses beyond the Project (a)
WBS Representations (b)
WBS Tools (c and d)

CHAPTER 3 QUESTIONS

1. Which of the following key project documents are created in the *Initiating* phase of a project?
(Select all that Apply)

   a. Project Charter
   b. Preliminary Project Scope Statement
   c. Product Scope Description
   d. Work Breakdown Structure

   Correct Answer

   c. Project Charter
   Preliminary Project Scope Statement

   Answer Reference

   Chapter: 3—Project Initiation and the WBS
   Section: Chapter Overview

2. Which foundational project management document provides the initial boundaries for the project's scope?

   a. Project Charter
   b. Preliminary Project Scope Statement
   c. Product Scope Description
   d. Work Breakdown Structure

   Correct Answer

   a. Project Charter

   Answer Reference

   Chapter: 3—Project Initiation and the WBS
   Section: Project Charter

3. Which foundational project management document is used to set the context for much of the *Planning* phase of the project?

   a. Project Charter
   b. Preliminary Project Scope Statement

c. Product Scope Description
d. Work Breakdown Structure

Correct Answer

b. Preliminary Project Scope Statement

Answer Reference

Chapter: 3—Project Initiation and the WBS
Section: Preliminary Project Scope Statement

4. Which foundational project management document includes information on how the final products, services or outcomes of the project will be measured?
   a. Project Charter
   b. Preliminary Project Scope Statement
   c. Product Scope Description
   d. Work Breakdown Structure

Correct Answer

b. Preliminary Project Scope Statement

Answer Reference

Chapter: 3—Project Initiation and the WBS
Section: Preliminary Project Scope Statement

5. Contracts should always be put in place before the project is fully defined?
   a. True
   b. False

Correct Answer

b. False

Answer Reference

Chapter: 3—Project Initiation and the WBS
Section: Contracts, Agreements, Statements of Work (SOW)

CHAPTER 4 QUESTIONS

1. Which of the following project documents describes the look and feel of the outputs of the project in narrative format?
   a. Project Charter

b. Preliminary Project Scope Statement
c. Product Scope Description
d. Work Breakdown Structure

Correct Answer

b. Product Scope Description

Answer Reference

Chapter: 4—Defining Scope through the WBS
Section: Product Scope Description

2. Quality Work Breakdown Structures include which of the following? (Select all that Apply)

a. Internal Deliverables
b. External Deliverables
c. Interim Deliverables
d. Ad hoc Deliverables

Correct Answer

a. Internal Deliverables
b. External Deliverables
c. Interim Deliverables

Answer Reference

Chapter: 4—Defining Scope through the WBS
Section: Work Breakdown Structure

3. Which of these WBS creation approaches involves first defining all of the detailed deliverables for the project?

a. Top-Down
b. Bottom-Up
c. WBS Standards
d. Templates

Correct Answer

b. Bottom-Up

Answer Reference

Chapter: 4—Defining Scope through the WBS
Section: Work Breakdown Structure

4. Quality Work Breakdown Structures must have at least ________________ levels of decomposition.
   a. One
   b. Two
   c. Three
   d. Four

   Correct Answer

   b. Two

   Answer Reference

   Chapter: 4—Defining Scope through the WBS
   Section: Beginning with the Elaborated WBS

5. Which approach to the management of projects allows for schedules, cost, resource and quality to be understood, aggregated, measured and monitored at both a specific deliverable and higher-level WBS element level?
   a. Activity-Based Management
   b. Task-Based Management
   c. Deliverable-Based Management
   d. Milestone-Based Management

   Correct Answer

   c. Deliverable-Based Management

   Answer Reference

   Chapter: 4—Defining Scope through the WBS
   Section: Deliverable-Based Management

CHAPTER 5 QUESTIONS

1. Work Package cost estimates should include which of the following? (Select all that Apply)
   a. Resource Costs
   b. Material Costs
   c. Quality Costs
   d. Risk Response Costs

e. Communications Costs
f. a and b only

Correct Answer

a. Resource Costs
b. Material Costs
c. Quality Costs
d. Risk Response Costs
e. Communications Costs

Answer Reference

Chapter: 5—The WBS in Procurement and Financial Planning
Section: Cost Estimating

2. Which scope-management-related deliverable can greatly aid in cost estimating?
   a. Project Charter
   b. Project Scope Statement
   c. Product Scope Description
   d. WBS Dictionary

Correct Answer

d. WBS Dictionary

Answer Reference

Chapter: 5—The WBS in Procurement and Financial Planning
Section: Cost Estimating

3. Which of the following are advantages for structuring cost budgets following the construct of the WBS?
   (Select all that Apply)
   a. The 100% Rule used in the creation of the WBS hierarchy also ensures that the cost budget will ultimately include 100% of the costs.
   b. The WBS elements in the hierarchical structure are used as the Control Account, thereby ensuring synchronization between how the work is defined and how it is performed and managed.
   c. The WBS provides for roll-up of costs, similar to deliverables roll-up in a WBS hierarchy.
   d. b and c Only

Correct Answer

a. The 100% Rule used in the creation of the WBS hierarchy also ensures that the cost budget will ultimately include 100% of the costs.
b. The WBS elements in the hierarchical structure are used as the Control Account, thereby ensuring synchronization between how the work is defined and how it is performed and managed.
c. The WBS provides for roll-up of costs, similar to deliverables roll-up in a WBS hierarchy.

Answer Reference

Chapter: 5—The WBS in Procurement and Financial Planning
Section: Cost Budgeting

4. What is a "management control point where the integration of scope, budget, actual cost, and schedule take place"?
   a. WBS Element
   b. Work Package
   c. Control Account
   d. None of the Above

Correct Answer

c. Control Account

Answer Reference

Chapter: 5—The WBS in Procurement and Financial Planning
Section: Cost Budgeting

5. What is a hierarchical breakdown of project cost components?
   a. Work Breakdown Structure
   b. Resource Breakdown Structure
   c. Organization Breakdown Structure
   d. Cost Breakdown Structure

Correct Answer

d. Cost Breakdown Structure

Answer Reference

Chapter: 5—The WBS in Procurement and Financial Planning
Section: Cost Breakdown Structure

CHAPTER 6 QUESTIONS

1. One sure path to success for the project manager is to apply tried and tested processes.
   a. True
   b. False

Correct Answer

a. True

Answer Reference

Chapter: 6—Quality, Risk, Resource and Communication Planning with the WBS
Section: Using Existing Templates and Processes

2. Which of the following can the Work Breakdown Structure be utilized for?
   (Select all that Apply)
   a. Envisioning work to be staffed
   b. Understanding the integration of components
   c. Imagining testing and verification steps
   d. a and b only

Correct Answer

a. Envisioning work to be staffed
b. Understanding the integration of components
c. Imagining testing and verification steps

Answer Reference

Chapter: 6—Quality, Risk, Resource and Communication Planning with the WBS
Section: The Whole is not Greater than the Sum of its Parts—It Equals Precisely 100% of the Sum of its Parts

3. Utilizing standard company processes for staffing individual project tasks is not beneficial to the project manager in organizations where processes are known and stable.

   a. True
   b. False

   Correct Answer

   b. False

   Answer Reference

   Chapter: 6—Quality, Risk, Resource and Communication Planning with the WBS
   Section: Using Existing Templates and Processes

4. The ______________ serves as the foundation for establishing entrance and exit criteria for the various stages of the project.

   a. Work Breakdown Structure
   b. Project Schedule
   c. Project Charter
   d. Scope Statement

   Correct Answer

   a. Work Breakdown Structure

   Answer Reference

   Chapter: 6—Quality, Risk, Resource and Communication Planning with the WBS
   Section: Process Considerations

5. The WBS can be considered a ______________ :
   (Select all that Apply)

   a. Communications Tool
   b. Scoping Tool
   c. Planning Tool
   d. Measuring Tool
   e. a, b and d

Correct Answer

a. Communications Tool
b. Scoping Tool
c. Planning Tool
d. Measuring Tool

Answer Reference

Chapter: 6—Quality, Risk, Resource and Communication Planning with the WBS
Section: Entire Chapter

CHAPTER 7 QUESTIONS

1. Which term describes the inputs, tools, techniques and outputs necessary to create the listing of activities that will be performed to produce desired project outcomes?
   a. Activity Definition
   b. Activity Sequencing
   c. Activity Estimation
   d. Work Breakdown Structure

   Correct Answer

   a. Activity Definition

   Answer Reference

   Chapter: 7—The WBS as a Starting Point for Schedule Development
   Section: Demystifying the Transition from the WBS to the Project Schedule

2. What term explains how the project's activities, milestones and approved changes are used as input to the activity sequencing process?
   a. Work Breakdown Structure
   b. Activity Definition
   c. Activity Estimation
   d. Activity Sequencing

   Correct Answer

   d. Activity Sequencing

Answer Reference

Chapter: 7—The WBS as a Starting Point for Schedule Development
Section: Demystifying the Transition from the WBS to the Project Schedule

3. Place the following deliverables in the proper sequential order of development by either filling in the blank table cells with the proper number or by re-ordering the deliverables.

Order	Deliverable
	Network Diagram
	Project Schedule
	WBS / WBS Dictionary

Correct Answer

Order	Deliverable
1	WBS / WBS Dictionary
2	Network Diagram
3	Project Schedule

Answer Reference

Chapter: 7—The WBS as a Starting Point for Schedule Development
Section: Demystifying the Transition from the WBS to the Project Schedule

4. Match each of the following elements to either a Work Breakdown Structure or a Project Schedule by filling in the table cell with the proper indicator—"Project Schedule" or "WBS".

Milestones	
Tasks	
Work Packages	
Activities	

Correct Answer

Milestones	Project Schedule
Tasks	Project Schedule
Work Packages	WBS
Activities	Project Schedule

Answer Reference

Chapter: 7—The WBS as a Starting Point for Schedule Development
Section: Entire Chapter

5. Fill in the blank with the appropriate words:
Inclusion as a dimension is used to show which elements are ________ ________ larger scope elements as well as clearly articulating which WBS elements are ________ ________ ________ the work of others.

Correct Answer

part of; not part of

Answer Reference

Chapter: 7—The WBS as a Starting Point for Schedule Development
Section: The Concept of Inclusion

CHAPTER 8 QUESTIONS

1. How often are the WBS and WBS Dictionary utilized to verify and validate that the appropriate resources are available and assigned to the project on what basis?
   a. Once
   b. Continually
   c. Never
   d. Sometimes

Correct Answer

b. Continually

Answer Reference

Chapter: 8—The WBS in Action
Section: Acquire Project Team

2. The WBS Dictionary is used to provide a detailed explanation of each ______________?
   a. Activity
   b. Task
   c. Deliverable
   d. Milestone

Correct Answer

c. Deliverable

Answer Reference

Chapter: 8—The WBS in Action
Section: Acquire Project Team

3. Place the following steps in order:

____	Update WBS Dictionary
____	Update Project Budget
____	Update Project Scope Statement
____	Update all Planning Documents
____	Update Work Breakdown Structure

Correct Answer

1	Update Scope Statement
2	Update Work Breakdown Structure
3	Update WBS Dictionary
4	Update all Planning Documents
5	Update Project Budget

Answer Reference

Chapter: 8—The WBS in Action
Section: Direct and Manage Project Execution and Integrated Change Management

4. Change Requests should be evaluated against this baseline in order to determine impact to scope.
   a. Project Management Plan
   b. Project Charter
   c. Project Schedule
   d. Work Breakdown Structure and Dictionary

Correct Answer

d. Work Breakdown Structure and Dictionary

Answer Reference

Chapter: 8—The WBS in Action
Section: Scope Management

5. Which of the following are not included in the WBS Dictionary?
   a. Acceptance Criteria
   b. Completion Criteria
   c. Quality Measures
   d. Test Cases

Correct Answer

d. Test Cases

Answer Reference

Chapter: 8—The WBS in Action
Section: Scope Verification

CHAPTER 9 QUESTIONS

1. Which of the following is *not* a source from which metrics used in project performance are derived?
   a. Issues Log
   b. WBS and WBS Dictionary

c. Project Schedule
d. Cost Management Plan

Correct Answer

a. Issues Log

Answer Reference

Chapter: 9—Ensuring Success through the WBS
Section: Project Performance Management

2. What is one of the drawbacks of the WBS?
   a. It has to be updated.
   b. It is hard to create.
   c. It can be perceived as one-dimensional.
   d. It can't be done on paper.

Correct Answer

c. It can be perceived as one-dimensional

Answer Reference

Chapter: 9—Ensuring Success through the WBS
Section: Scope

3. Before a Project Schedule can be used for Performance Reporting, what final step must be taken?
   a. Create the Schedule
   b. Update the Schedule
   c. Add the Schedule to a Project Scheduling Tool
   d. Baseline the Schedule

Correct Answer

d. Baseline the Schedule

Answer Reference

Chapter: 9—Ensuring Success through the WBS
Section: Schedule

4. Which project management analysis technique enables the Project Manager to predict the cost and delivery date of the project based on its performance over time?
   a. Cost Management
   b. Work Breakdown Structures

c. Earned Value Management
d. Project Scoping

Correct Answer

c. Earned Value Management

Answer Reference

Chapter: 9—Ensuring Success through the WBS
Section: Schedule

5. Which is a recommended method for organizing project data for presentation to stakeholders?
   a. Grouped alphabetically
   b. Grouped chronologically (by schedule)
   c. Grouped by Cost Account
   d. Grouped by Work Breakdown Structure hierarchy

Correct Answer

d. Grouped by Work Breakdown Structure hierarchy

Answer Reference

Chapter: 9—Ensuring Success through the WBS
Section: Planned vs. Actual

CHAPTER 10 QUESTIONS

1. The WBS and WBS Dictionary are vitally important during *all* phases of the project lifestyle?
   a. False
   b. True

Correct Answer

b. True

Answer Reference

Chapter: 10—Verifying Project Closeout with the WBS
Section: Chapter Summary

2. Which of the following are *not* utilized during negotiations about acceptability of the delivered product(s)?
   a. WBS
   b. Contracts / Agreements

c. WBS Dictionary
d. Risk Register

Correct Answer

d. Risk Register

Answer Reference

Chapter: 10—Verifying Project Closeout with the WBS
Section: Acceptance / Turnover / Support / Maintenance

3. Changes to the baselined Work Breakdown Structure should be made through which project process?
   a. Project Planning
   b. Change Management
   c. Scoping
   d. Project Closeout

Correct Answer

b. Change Management

Answer Reference

Chapter: 10—Verifying Project Closeout with the WBS
Section: Project Closeout

4. What is the first step in project Closing activities?
   a. Celebrate
   b. Contract Closure
   c. Initiate a Post-Project Review
   d. Verify all Deliverables have been completed
   e. Update all documentation to record and reflect final results

Correct Answer

d. Verify all deliverables have been completed

Answer Reference

Chapter: 10—Verifying Project Closeout with the WBS
Section: Project Closeout, Contract Closure

5. The Work Breakdown Structure can be utilized as basis for post project review?
   a. True
   b. False

Correct Answer

a. True

Answer Reference

Chapter: 10—Verifying Project Closeout with the WBS
Section: Project Closeout

Index

D

E

F

G

H

Q

R

CPSIA information can be obtained
at www.ICGtesting.com
Printed in the USA
BVHW040858230320
575590BV00030B/60